Springer Series in Operations Research and Financial Engineering

Series Editors:
Thomas V. Mikosch
Sidney I. Resnick
Stephen M. Robinson

For further volumes:
http://www.springer.com/series/3182

Sergey Foss • Dmitry Korshunov • Stan Zachary

An Introduction to Heavy-Tailed and Subexponential Distributions

Second Edition

 Springer

Sergey Foss
Department of Actuarial Mathematics
Heriot-Watt University
Riccarton, Edinburgh, UK

Dmitry Korshunov
Sobolev Institute of Mathematics
Novosibirsk, Russia

Stan Zachary
Department of Actuarial Mathematics
Heriot-Watt University
Riccarton, Edinburgh, UK

ISSN 1431-8598
ISBN 978-1-4899-8832-4   ISBN 978-1-4614-7101-1 (eBook)
DOI 10.1007/978-1-4614-7101-1
Springer New York Heidelberg Dordrecht London

Mathematics Subject Classification (2010): 60E99, 62E20, 60F10, 60G50

Printed on acid-free paper

Springer is part of Springer Science+Business Media (www.springer.com)

Preface to the First Edition

This text studies heavy-tailed distributions in probability theory, and especially convolutions of such distributions. The main goal is to provide a complete and comprehensive introduction to the theory of long-tailed and subexponential distributions which includes many novel elements and, in particular, is based on the regular use of the principle of a single big jump. Much of the material appears for the first time in text form, including:
- The establishment of new relations between known classes of subexponential distributions and the introduction of important new classes
- The development of some important new concepts, including those of h-insensitivity and local subexponentiality
- The presentation of new and direct probabilistic proofs of known asymptotic results

A number of recent textbooks and monographs contain some elements of the present theory, notably those by S. Asmussen [1, 2], P. Embrechts, C. Klüppelberg, and T. Mikosch [24], T. Rolski, H. Schmidli, V. Schmidt, and J. Teugels [47], and A. Borovkov and K. Borovkov [11]. Further, the monograph by N. Bingham, C. Goldie, and J. Teugels [9] comprehensively develops the theory of regularly varying functions and distributions; the latter form an important subclass of the subexponential distributions. We have been influenced by these books and by further contacts with their authors.

Chapters 2 and 3 of the present monograph deal comprehensively with the properties of heavy-tailed, long-tailed and subexponential distributions, and give applications to random sums. Chapter 4 develops concepts of local subexponentiality and gives further applications. Finally, Chap. 5 studies the distribution of the maximum of a random walk with negative drift and heavy-tailed increments; notably it contains new and short probabilistic proofs for the tail asymptotics of this distribution for both finite and infinite time horizons. The study of heavy-tailed distributions in more general probability models—for example, Markov-modulated models, those with dependencies, and continuous-time models—is postponed until such future date as the authors may again find some spare time. Nevertheless, the same basic principles apply there as are developed in the present text.

The authors gratefully acknowledge a fruitful collaboration on heavy-tails issues with their co-authors: Søren Asmussen, François Baccelli, Aleksandr Borovkov, Onno Boxma, Denis Denisov, Takis Konstantopoulos, Marc Lelarge, Andrew Richards, and Volker Schmidt.

We are thankful to many colleagues, in addition to mentioned above, for helpful discussions and contributions, notably to Vsevolod Shneer and Bert Zwart. We thank Sergei Fedotov for pointing out links to analogous problems in Statistical Physics.

We are also very grateful both to Thomas Mikosch and to the staff of Springer for their suggestions and assistance in publishing this text.

This book was mostly written while the authors worked, together and individually, in Edinburgh and Novosibirsk; we thank our home institutions, Heriot-Watt University and the Sobolev Institute of Mathematics. A first version of this manuscript was finished while the authors stayed at the Mathematisches Forschungsinstitut Oberwolfach, under the Research in Pairs programme from March 23 to April 5, 2008; we thank the Institute for its great hospitality and support. The final version was prepared in Cambridge during our stay at the Isaac Newton Institute for Mathematical Sciences under the framework of the programme Stochastic Processes in Communication Sciences, January–June, 2010.

A list of errata and notes on further developments to this manuscript will be maintained at https://sites.google.com/site/ithtsd/ https://sites.google.com/site/ithtsd/.

Edinburgh   Sergey Foss
Novosibirsk   Dmitry Korshunov
Oberwolfach and Cambridge   Stan Zachary

Preface

This is an extended and corrected version of the First Edition. The major changes are:
- Chapters 2 through 5 are now appended by lists of problems and exercises. We also provide answers and a number of solutions.
- Chapter 5 includes three new sections on applications, to queueing theory, to risk, and to branching processes, and a new section describing time to exceed a high level by a random walk and its location around that time.
- Sections 5.1, 5.2 and 5.9 are extended.

Sergey Foss

August 2012　Dmitry Korshunov

Contents

Notation and Conventions

Intervals	(x,y) is an open, $[x,y]$ a closed interval; half-open intervals are denoted by $(x,y]$ and $[x,y)$.
Integrals	$\int_x^y$ is the integral over the interval $(x,y]$.
$\mathbb{R}, \mathbb{R}^+, \mathbb{R}^s$	Stand for the real line, the positive real half-line $[0,\infty)$, and s-dimensional Cartesian space.
$\mathbb{Z}, \mathbb{Z}^+$	Stand for the set of integers and for the set $\{0,1,2,\ldots\}$.
$\mathbb{I}(A)$	Stands for the indicator function of A, i.e. $\mathbb{I}(A) = 1$ if A holds and $\mathbb{I}(A) = 0$ otherwise.
$O, o,$ *and* $\sim$	Let u and v depend on a parameter x which tends, say, to infinity. Assuming that v is positive we write

$$u = O(v) \ \text{ if } \ \limsup_{x \to \infty} |u|/v < \infty$$
$$u = o(v) \ \text{ if } \ u/v \to 0 \text{ as } x \to \infty$$
$$u \sim v \ \text{ if } \ u/v \to 1 \text{ as } x \to \infty.$$

$\mathbb{P}\{B\}$	Stands for the probability (on some appropriate space) of the event B.
$\mathbb{P}\{B|A\}$	Stands for the conditional probability of the event B given A, i.e., for the ratio $\mathbb{P}\{BA\}/\mathbb{P}\{A\}$.
$\mathbb{E}\xi$	Stands for the mean of the random variable ξ.
$\mathbb{E}\{\xi;B\}$	Stands for the mean of ξ over the event B, i.e., for $\mathbb{E}\xi\mathbb{I}(B)$.
$F * G$	Stands for the convolution of the distributions F and G.
F^{*n}	Stands for the n-fold convolution of the distribution F with itself.
ξ^+, F^+	For any random variable ξ on $\mathbb{R}$ with distribution F, the random variable $\xi^+ = \max(\xi,0)$ and F^+ denotes its distribution.
$:= (=:)$	The quantity on the left (right) is defined to be equal to the quantity on the right (left).
$\square$	Indicates the end of a proof.

Chapter 1
Introduction

Heavy-tailed distributions (probability measures) play a major role in the analysis of many stochastic systems. For example, they are frequently necessary to accurately model inputs to computer and communications networks, they are an essential component of the description of many risk processes, they occur naturally in models of epidemiological spread, and there is much statistical evidence for their appropriateness in physics, geoscience and economics. Important examples are Pareto distributions (and other essentially power-law distributions), lognormal distributions, and Weibull distributions (with shape parameter less than 1). Indeed most heavy-tailed distributions used in practice belong to one of these families, which are defined, along with others, in Chap. 2. We also consider the Weibull distribution at the end of this chapter.

Since the inputs to systems such as those described above are frequently cumulative in their effects, the analysis of the corresponding models typically features convolutions of heavy-tailed distributions. For a satisfactory theory it is necessary that these distributions possess certain regularity conditions. From the point of view of applications practically all heavy-tailed distributions may be considered to be long-tailed, and indeed to possess the stronger property of subexponentiality (see below for definitions).

In this monograph we study convolutions of long-tailed and subexponential distributions on the real line. Our aim is to prove some important new results, and to do so through a simple, coherent and systematic approach. It turns out that all the standard properties of such convolutions are then obtained as easy consequences of these results. Thus we also hope to provide further insight into these properties, and to dispel some of the mystery which still seems to surround the phenomenon of subexponentiality in particular.

We define the *tail function* $\overline{F}$ of a distribution F on $\mathbb{R}$ to be given by $\overline{F}(x) = F(x, \infty)$ for all x. We describe as a *tail property* of F any property which depends only on $\{\overline{F}(x) : x \geq x_0\}$ for any (finite) x_0. We further say that F has *right-unbounded support* if $\overline{F}(x) > 0$ for all x.

S. Foss et al., *An Introduction to Heavy-Tailed and Subexponential Distributions*, Springer Series in Operations Research and Financial Engineering, DOI 10.1007/978-1-4614-7101-1_1, © Springer Science+Business Media New York 2013

Heavy-Tailed Distributions

A distribution F on $\mathbb{R}$ is said to be *(right-) heavy-tailed* if

$$\int_{-\infty}^{\infty} e^{\lambda x} F(dx) = \infty \quad \text{for all } \lambda > 0, \tag{1.1}$$

that is, if and only if F fails to possess any positive exponential moment. Otherwise F is said to be *light-tailed*. We shall show in Chap. 2 that the distribution F is heavy-tailed if and only if its tail function $\overline{F}$ fails to be bounded by any exponentially decreasing function.

It follows that for a distribution F to be heavy-tailed is a tail property of F, and of course that any heavy-tailed distribution has right-unbounded support.

We mention briefly at this point the connection with hazard rates. Let F be a distribution on $\mathbb{R}$ which is absolutely continuous with density f with respect to Lebesgue measure. Such a distribution is often characterised in terms of its *hazard rate* $r(x) = f(x)/\overline{F}(x)$, most naturally in the case where F is concentrated on the positive half-line $\mathbb{R}^+$. We then have

$$\overline{F}(x) = \exp\left(-\int_{-\infty}^{x} r(y)\, dy\right).$$

It follows easily from (1.1) that if $\lim_{x \to \infty} r(x) = 0$ then the distribution F is heavy-tailed, whereas if $\liminf_{x \to \infty} r(x) > 0$ then F fails to be heavy-tailed (indeed the integral in (1.1) is finite for any λ such that $\liminf_{x \to \infty} r(x) > \lambda$). In the final case where $\liminf_{x \to \infty} r(x) = 0$ but in which the limit itself fails to exist then both possibilities for F exist.

Long-Tailed Distributions

A distribution F on $\mathbb{R}$ is said to be *long-tailed* if F has right-unbounded support and, for any fixed $y > 0$,

$$\frac{\overline{F}(x+y)}{\overline{F}(x)} \to 1 \qquad \text{as } x \to \infty. \tag{1.2}$$

Clearly to be long-tailed is again a tail property of a distribution. Further, it is fairly easy to see that a long-tailed distribution is also heavy-tailed. However, the condition (1.2) implies a degree of smoothness in the tail function $\overline{F}$ which is not possessed by every heavy-tailed distribution.

Subexponential Distributions

In order to make good progress with heavy-tailed distributions, we require a slightly stronger regularity condition than the requirement that such a distribution be long-tailed. This will turn out to be satisfied by all heavy-tailed distributions likely to be encountered in practice.

We consider first distributions on the positive half-line $\mathbb{R}^+$. Let F be any distribution on $\mathbb{R}^+$ and let $\xi_1, \ldots, \xi_n$ be independent random variables with the common distribution F. Then

$$\mathbb{P}\{\xi_1 + \ldots + \xi_n > x\} \geq \mathbb{P}\{\max(\xi_1, \ldots, \xi_n) > x\}$$
$$= 1 - F^n(x)$$
$$\sim n\overline{F}(x) \quad \text{as } x \to \infty. \tag{1.3}$$

(Here and throughout we use "$\sim$" to mean that the ratio of the quantities on either side of this symbol converges to one; we further frequently omit, especially in proofs, the qualifier "as $x \to \infty$", as unless otherwise indicated all our limits will be of this form.)

Taking $n = 2$ it follows in particular that

$$\liminf_{x \to \infty} \frac{\overline{F * F}(x)}{\overline{F}(x)} \geq 2, \tag{1.4}$$

where as usual, for any two distributions F and G, by $F * G$ we denote their *convolution*, i.e. the distribution of the random variable $\xi + \eta$ where the random variables ξ and η are independent with distributions F and G.

A considerably deeper result (proved in Chap. 2) than the above inequality (1.4) is that if F is *heavy-tailed* then the relation (1.4) holds with *equality*. (We remark that there are also examples of light-tailed distributions on $\mathbb{R}^+$ for which (1.4) holds with equality.) The distribution F on $\mathbb{R}^+$ is said to be *subexponential* if

$$\lim_{x \to \infty} \frac{\overline{F * F}(x)}{\overline{F}(x)} = 2. \tag{1.5}$$

It turns out that the above condition is now sufficient to ensure that F is heavy-tailed—and indeed that F is long-tailed. Thus a distribution F on $\mathbb{R}^+$ is subexponential if and only if is heavy-tailed and sufficiently regular that the limit on the left side of (1.5) exists; this limit is then equal to 2. It is therefore not surprising that the various examples of heavy-tailed distributions on $\mathbb{R}^+$ mentioned at the start of this chapter all turn out to be subexponential. Indeed those heavy-tailed distributions which do not possess this property are all distinctly pathological in character.

We shall see that subexponentiality as defined above is also a *tail property* of the distribution F. Inductive arguments (see Chap. 3) now show that if a distribution F on $\mathbb{R}^+$ is subexponential then the relation (1.5) generalises to

$$\lim_{x \to \infty} \frac{\overline{F^{*n}}(x)}{\overline{F}(x)} = n \quad \text{for all integer } n \geq 1$$

(where F^{*n} denotes the n-fold convolution of the distribution F with itself). It follows from this and from the argument leading to (1.3) that subexponentiality of F is equivalent to the requirement that

$$\mathbb{P}\{\max(\xi_1,\ldots,\xi_n) > x\} \sim \mathbb{P}\{\xi_1 + \cdots + \xi_n > x\} \quad \text{as } x \to \infty. \tag{1.6}$$

The interpretation of the condition (1.6) is that the only significant way in which the sum $\xi_1 + \cdots + \xi_n$ can exceed some large value x is that the maximum of one of the individual random variables $\xi_1, \ldots, \xi_n$ also exceeds x. This is the *principle of a single big jump* which underlies the probabilistic behaviour of sums of independent subexponential random variables.

Since subexponentiality is a tail property of a distribution, it is natural, and important for many applications, to extend the concept to a distribution F on the entire real line $\mathbb{R}$. This may be done *either* by requiring that F has the same tail as that of a subexponential distribution on $\mathbb{R}^+$ (it is natural to consider the distribution F^+ given by $F^+(x) = F(x)$ for $x \geq 0$ and $F^+(x) = 0$ for $x < 0$) *or*, equivalently as it turns out, by requiring that F is long-tailed and again satisfies the condition (1.5)—the latter condition on its own no longer being sufficient for the subexponentiality of F. We explore these matters further in Chap. 3.

We develop also similar concepts of subexponentiality for local probabilities and for densities (see Chap. 4).

Further examples of heavy-tailed distributions which are of use in practical applications, e.g. the modelling of insurance claim sizes, are given by Embrechts, Klüppelberg, and Mikosch [24]. These, and the examples mentioned above, are all well-behaved in a manner we shall shortly make precise. However, mathematically there is a whole range of further possible distributions, and one of our aims is to provide a firm basis for excluding those which are in some sense pathological and to study the properties of those which remain.

Example: The Weibull Distribution

In order to understand better the typical behaviour of heavy-tailed distributions, that is, the single big jump phenomenon—as opposed to the behaviour of distributions which are light-tailed—we study the *Weibull distribution* F_α given by its tail function

$$\overline{F}_\alpha(x) = e^{-x^\alpha}, \quad x \geq 0, \tag{1.7}$$

and hence density $f_\alpha(x) = \alpha x^{\alpha-1} e^{-x^\alpha}$, $x \geq 0$, for some *shape* parameter $\alpha > 0$. This is a heavy-tailed distribution if and only if $\alpha < 1$, and is then sometimes called a *stretched exponential distribution*—notably in physics. We refer to Hallinan [29] for a historical review of Weibull distribution. Note that in the case $\alpha = 1$ we have the *exponential distribution*. All moments of the Weibull distribution are finite.

In practice, this class of distributions is motivated in part by the fact that the large deviations of multiplicative processes are usually Weibull-distributed, see Frisch and Sornette [28]. For example, consider n independent random variables

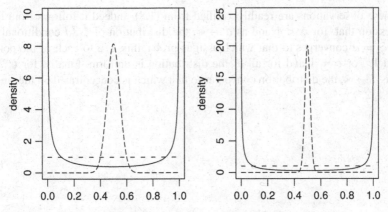

Fig. 1.1 Density of ξ_1/d conditional on $\xi_1 + \xi_2 = d$, for $d = 10$ (*left panel*) and $d = 25$ (*right panel*), and for $\alpha = 0.5$ (*solid line*), $\alpha = 1$ (*short-dashed line*) and $\alpha = 2$ (*long-dashed line*)

$\xi_1, \ldots, \xi_n$ with a common light-tailed distribution F such that $\overline{F}(x) \sim e^{-cx^\gamma}$ with $\gamma \geq 1$, for instance an exponential or normal distribution. Then the tail of the distribution of their product $\xi_1 \ldots \xi_n$ possesses the following lower bound:

$$\mathbb{P}\{\xi_1 \ldots \xi_n > x\} \geq \mathbb{P}\{\xi_1 > \sqrt[n]{x}, \ldots, \xi_n > \sqrt[n]{x}\}$$
$$= (\overline{F}(\sqrt[n]{x}))^n \sim e^{-cnx^{\gamma/n}},$$

and the shape parameter of this Weibull distribution is less than 1 when $n > \gamma$; for the exponential distribution the product $\xi_1 \xi_2$ is heavy-tailed, while for the normal distribution the triple product $\xi_1 \xi_2 \xi_3$ is heavy tailed. Many examples of why the Weibull distribution is valuable in describing different phenomena in nature and in economics may be found in Laherrère and Sornette [39], Malevergne and Sornette [42], Sornette [51] and Metzler and Klafter [43].

Now let ξ_1 and ξ_2 be independent random variables with common Weibull distribution function F_α as given by (1.7). We consider the distribution of the random variable ξ_1/d conditional on the sum $\xi_1 + \xi_2 = d$ for varying values of d and the shape parameter α. This conditional distribution has density $g_{\alpha,d}$ where

$$g_{\alpha,d}(z) = c[z(1-z)]^{\alpha-1} e^{-d^\alpha(z^\alpha + (1-z)^\alpha)}, \qquad (1.8)$$

for the appropriate normalising constant c. Clearly this conditional density is symmetric about $1/2$. The left panel of Fig. 1.1 plots the density for $d = 10$ and for each of the three cases $\alpha = 0.5$, $\alpha = 1$, and $\alpha = 2$, while the right panel plots the density for $d = 25$ and for each of the same three values of α. We see that in the heavy-tailed case $\alpha = 0.5$, conditional on the fixed value d of the sum $\xi_1 + \xi_2$, the value of ξ_1/d tends to be either close to 0 or close to 1; further this effect is more pronounced for the larger value of d. For the case $\alpha = 1$ and for any value of d, the above conditional density is uniform. For the case $\alpha = 2$, we see that the conditional density of ξ_1/d is concentrated in a neighbourhood of $1/2$, and that again this concentration is more pronounced for the larger value of d.

These observations are readily verified from (1.8). Indeed it follows from that expression that, for $\alpha < 1$ and as $d \to \infty$, the distribution of ξ_1/d conditional on $\xi_1 + \xi_2 = d$ converges to that which assigns probability $1/2$ to each of the points 0 and 1. For $\alpha = 1$ and for all d, the distribution is uniform. Finally, for $\alpha > 1$ and as $d \to \infty$, the distribution converges to that which is concentrated on the single point $1/2$.

Chapter 2
Heavy-Tailed and Long-Tailed Distributions

In this chapter we are interested in *(right-) tail properties* of distributions, i.e. in properties of a distribution which, for any x, depend only on the restriction of the distribution to (x, ∞). More generally it is helpful to consider tail properties of functions.

Recall that for any distribution F on $\mathbb{R}$ we define the *tail function* $\overline{F}$ by

$$\overline{F}(x) = F(x, \infty), \quad x \in \mathbb{R}.$$

We start with characteristic properties of heavy-tailed distributions, i.e., of distributions all of whose positive exponential moments are infinite. The main result here concerns lower limits for convolution tails, see Sect. 2.3.

Following this we study different properties of long-tailed distributions, i.e., of distributions whose tails are asymptotically self-similar under shifting by a constant. Of particular interest are convolutions of long-tailed distributions. Our approach is based on a simple decomposition for such convolutions and on the concept of "h-insensitivity" for a long-tailed distribution with respect to some (slowly) increasing function h. In Sect. 2.8, we present useful characterisations of h-insensitive distributions.

2.1 Heavy-Tailed Distributions

The usage of the term "heavy-tailed distribution" varies according to the area of interest but is frequently taken to correspond to an absence of (positive) exponential moments. In the following definitions—which, for completeness here, repeat some of those made in the Introduction—we follow this tradition.

Definition 2.1. A distribution F on $\mathbb{R}$ is said to have *right-unbounded support* if $\overline{F}(x) > 0$ for all x.

Definition 2.2. We define a *distribution F* to be (right-) *heavy-tailed* if and only if

S. Foss et al., *An Introduction to Heavy-Tailed and Subexponential Distributions*,
Springer Series in Operations Research and Financial Engineering,
DOI 10.1007/978-1-4614-7101-1_2, © Springer Science+Business Media New York 2013

$$\int_{\mathbb{R}} e^{\lambda x} F(dx) = \infty \quad \text{for all } \lambda > 0. \tag{2.1}$$

It will follow from Theorem 2.6 that to be heavy-tailed is indeed a tail property of a distribution. As a counterpart we give also the following definition.

Definition 2.3. A *distribution F* is called *light-tailed* if and only if

$$\int_{\mathbb{R}} e^{\lambda x} F(dx) < \infty \quad \text{for some } \lambda > 0, \tag{2.2}$$

i.e. if and only if it fails to be heavy-tailed.

Clearly, for any light-tailed distribution F on the positive half-line $\mathbb{R}^+ = [0, \infty)$, all moments are finite, i.e., $\int_0^\infty x^k F(dx) < \infty$ for all $k > 0$.

We shall say that a non-negative *function* (usually tending to zero) is *heavy-tailed* if it fails to be bounded by a decreasing exponential function. More precisely we make the following definition.

Definition 2.4. We define a *function* $f \geq 0$ to be *heavy-tailed* if and only if

$$\limsup_{x \to \infty} f(x) e^{\lambda x} = \infty \quad \text{for all } \lambda > 0. \tag{2.3}$$

For a function to be heavy-tailed is clearly a tail-property of that function. Theorem 2.6 shows in particular that a *distribution* is heavy-tailed if and only if its tail function is a heavy-tailed function. First we make the following definition.

Definition 2.5. For any distribution F, the function $R(x) := -\ln \overline{F}(x)$ is called the *hazard function* of the distribution. If the hazard function is differentiable, then its derivative $r(x) = R'(x)$ is called the *hazard rate*.

The hazard rate, when it exists, has the usual interpretation discussed in the Introduction.

Theorem 2.6. *For any distribution F the following assertions are equivalent:*
  (i) *F is a heavy-tailed distribution.*
  (ii) *The function $\overline{F}$ is heavy-tailed.*
  (iii) *The corresponding hazard function R satisfies $\liminf_{x \to \infty} R(x)/x = 0$.*
  (iv) *For some (any) fixed $T > 0$, the function $F(x, x+T]$ is heavy-tailed.*

Proof. (i)$\Rightarrow$(iv). Suppose that the function $F(x, x+T]$ is not heavy-tailed. Then

$$c := \sup_{x \in \mathbb{R}} F(x, x+T] e^{\lambda' x} < \infty \quad \text{for some } \lambda' > 0,$$

and, therefore, for all $\lambda < \lambda'$

$$\int_0^\infty e^{\lambda x} F(dx) \leq \sum_{n=0}^\infty e^{\lambda(n+1)T} F(nT, nT+T]$$

$$\leq c \sum_{n=0}^\infty e^{\lambda(n+1)T} e^{-\lambda' nT} = c e^{\lambda T} \sum_{n=0}^\infty e^{(\lambda-\lambda')nT} < \infty.$$

It follows that the integral defined in (2.1) is finite for all $\lambda \in (0, \lambda')$, which implies that the distribution F cannot be heavy-tailed. The required implication now follows.

(iv)$\Rightarrow$(ii). This implication follows from the inequality $\overline{F}(x) \geq F(x, x+T]$.

(ii)$\Rightarrow$(iii). Suppose that, on the contrary, "lim inf" in (iii) is (strictly) positive. Then there exist $x_0 > 0$ and $\varepsilon > 0$ such that $R(x) \geq \varepsilon x$ for all $x \geq x_0$ which implies that $\overline{F}(x) \leq e^{-\varepsilon x}$ in contradiction of (ii).

(iii)$\Rightarrow$(i). Suppose that, on the contrary, F is light-tailed. It then follows from (2.2) (e.g., by the exponential Chebyshev inequality) that, for some $\lambda > 0$ and $c > 0$, we have $\overline{F}(x) \leq c e^{-\lambda x}$ for all x. This implies that $\liminf_{x \to \infty} R(x)/x \geq \lambda$ which contradicts (iii). $\hspace{1em}\square$

Lemma 2.7. *Let the distribution F be absolutely continuous with density function f. Suppose that the distribution F is heavy-tailed. Then the function $f(x)$ is heavy-tailed also.*

Proof. Suppose that $f(x)$ is not heavy-tailed; then there exist $\lambda' > 0$ and x_0 such that

$$c := \sup_{x > x_0} f(x) e^{\lambda' x} < \infty,$$

and, therefore, for all $\lambda \in (0, \lambda')$

$$\int_{\mathbb{R}} e^{\lambda x} F(dx) \leq e^{\lambda x_0} + c \int_{x_0}^{\infty} e^{\lambda x} e^{-\lambda' x} dx < \infty.$$

It follows that the integral defined in (2.1) is finite for all λ such that $0 < \lambda < \lambda'$, which contradicts heavy-tailedness of the distribution F. $\hspace{1em}\square$

We give an example to show that the converse assertion is not in general true. Consider the following piecewise continuous density function:

$$f(x) = \sum_{n=1}^{\infty} \mathbb{I}\{x \in [n, n + 2^{-n}]\}.$$

We have $\limsup_{x \to \infty} f(x) e^{\lambda x} = \infty$ for all $\lambda > 0$, so that f is heavy-tailed. On the other hand, for all $\lambda \in (0, \ln 2)$,

$$\int_0^{\infty} e^{\lambda x} f(x) dx < \sum_{n=1}^{\infty} e^{\lambda(n+2^{-n})} 2^{-n} = \sum_{n=1}^{\infty} e^{\lambda(n+2^{-n}) - n \ln 2} < \infty,$$

so that F is light-tailed.

For lattice distributions we have the following result.

Lemma 2.8. *Let F be a distribution on some lattice $\{a + hn, n \in \mathbb{Z}\}$, $a \in \mathbb{R}$, $h > 0$, with probabilities $F\{a + hn\} = p_n$. Then F is heavy-tailed if and only if the sequence $\{p_n\}$ is heavy-tailed, i.e.,*

$$\limsup_{n \to \infty} p_n e^{\lambda n} = \infty \quad \text{for all } \lambda > 0. \hspace{2em} (2.4)$$

Proof. The result follows from Theorem 2.6 with $T = h$. $\square$

Examples of Heavy-Tailed Distributions

We conclude this section with a number of examples.

- The *Pareto distribution* on $\mathbb{R}^+$. This has tail function $\overline{F}$ given by

$$\overline{F}(x) = \left(\frac{\kappa}{x+\kappa} \right)^{\alpha}$$

  for some scale parameter $\kappa > 0$ and shape parameter $\alpha > 0$. Clearly we have $\overline{F}(x) \sim (x/\kappa)^{-\alpha}$ as $x \to \infty$, and for this reason the Pareto distributions are sometimes referred to as the *power law distributions*. The Pareto distribution has all moments of order $\gamma < \alpha$ finite, while all moments of order $\gamma \geq \alpha$ are infinite.
- The *Burr distribution* on $\mathbb{R}^+$. This has tail function $\overline{F}$ given by

$$\overline{F}(x) = \left(\frac{\kappa}{x^{\tau} + \kappa} \right)^{\alpha}$$

  for parameters $\alpha, \kappa, \tau > 0$. We have $\overline{F}(x) \sim \kappa^{\alpha} x^{-\tau \alpha}$ as $x \to \infty$; thus the Burr distribution is similar in its tail to the Pareto distribution, of which it is otherwise a generalisation. All moments of order $\gamma < \alpha \tau$ are finite, while those of order $\gamma \geq \alpha \tau$ are infinite.
- The *Cauchy distribution* on $\mathbb{R}$. This is most easily given by its density function f where

$$f(x) = \frac{\kappa}{\pi((x-a)^2 + \kappa^2)}$$

  for some scale parameter $\kappa > 0$ and position parameter $a \in \mathbb{R}$. All moments of order $\gamma < 1$ are finite, while those of order $\gamma \geq 1$ are infinite.
- The *lognormal distribution* on $\mathbb{R}^+$. This is again most easily given by its density function f, where

$$f(x) = \frac{1}{\sqrt{2\pi}\sigma x} \exp\left(-\frac{(\log x - \mu)^2}{2\sigma^2} \right)$$

  for parameters μ and $\sigma > 0$. All moments of the lognormal distribution are finite. Note that a (positive) random variable ξ has a lognormal distribution with parameters μ and σ if and only if $\log \xi$ has a *normal distribution* with mean μ and variance σ^2. For this reason the distribution is natural in many applications.
- The *Weibull distribution* on $\mathbb{R}^+$. This has tail function $\overline{F}$ given by

$$\overline{F}(x) = e^{-(x/\kappa)^{\alpha}}$$

for some scale parameter $\kappa > 0$ and shape parameter $\alpha > 0$. This is a heavy-tailed distribution if and only if $\alpha < 1$. Note that in the case $\alpha = 1$ we have the *exponential distribution*. All moments of the Weibull distribution are finite.

Another useful class of heavy-tailed distributions is that of dominated-varying distributions. We say that F is a *dominated-varying distribution* (and write $F \in \mathcal{D}$) if there exists $c > 0$ such that

$$\overline{F}(2x) \geq c\overline{F}(x) \quad \text{for all } x.$$

Any intermediate regularly varying distribution (see Sect. 2.8) belongs to $\mathcal{D}$. Other examples may be constructed using the following scheme. Let G be a distribution with a regularly varying tail (again see Sect. 2.8). Then a distribution F belongs to the class $\mathcal{D}$, provided $c_1\overline{G}_1(x) \leq \overline{F}(x) \leq c_2\overline{G}(x)$ for some $0 < c_1 < c_2 < \infty$ and for all sufficiently large x.

2.2 Characterisation of Heavy-Tailed Distributions in Terms of Generalised Moments

A major objective of this and the succeeding section is to establish the important, if somewhat analytical, result referred to in the Introduction that for a heavy-tailed distribution F on $\mathbb{R}^+$ we have $\liminf_{x \to \infty} \overline{F*F}(x)/\overline{F}(x) = 2$. This is Theorem 2.12. As remarked earlier, it will then follow (see Chap. 3) that the subexponentiality of a distribution F on $\mathbb{R}^+$ is then *equivalent* to heavy-tailedness plus the reasonable regularity requirement that the limit as $x \to \infty$ of $\overline{F*F}(x)/\overline{F}(x)$ should exist.

In this section we therefore consider an important (and again quite analytical) characterisation of heavy-tailed distributions on $\mathbb{R}^+$, which is both of interest in itself and essential to the consideration of convolutions in the following section. In very approximate terms, for any such distribution we seek the existence of a monotone concave function h such that the function $e^{-h(\cdot)}$ characterises the tail of the distribution.

If a distribution F on the positive half-line $\mathbb{R}^+$ is such that not all of its moments are finite, i.e., $\int_0^\infty x^k F(dx) = \infty$ for some k, then F is heavy-tailed. In this case we can find such $k \geq 1$ that the kth moment is infinite, while the $(k-1)$th moment is finite. That is

$$\int_0^\infty xe^{(k-1)\ln x} F(dx) = \infty \quad \text{and} \quad \int_0^\infty e^{(k-1)\ln x} F(dx) < \infty. \tag{2.5}$$

Note that here the power of the exponent is a concave function. This observation can be generalised onto the whole class of heavy-tailed distributions as follows.

Theorem 2.9. *Let $\xi \geq 0$ be a random variable with a heavy-tailed distribution. Let the function $g(x)$ be such that $g(x) \to \infty$ as $x \to \infty$. Then there exists a monotone concave function $h : \mathbb{R}^+ \to \mathbb{R}^+$ such that $h(x) = o(x)$ as $x \to \infty$, $\mathbb{E}e^{h(\xi)} < \infty$, and $\mathbb{E}e^{h(\xi)+g(\xi)} = \infty$.*

Now (2.5) is a particular example of the latter theorem with $g(x) = \ln x$. In this case, if not all moments of ξ are finite, the concave function $h(x)$ may be taken as $(k-1)\ln x$ for k as defined above. However, Theorem 2.9 is considerably sharper: it guarantees the existence of a concave function h for any function g (such that $g(x) \to \infty$ as $x \to \infty$), which may be taken as slowly increasing as we please.

As a further example, note that if ξ has a Weibull distribution with tail function $\overline{F}(x) = e^{-x^\alpha}$, $\alpha \in (0,1)$, and if $g(x) = \ln x$, then one can choose $h(x) = (x+c)^\alpha - \ln(x+c)$, with $c > 0$ sufficiently large.

Note also that Theorem 2.9 provides a characteristic property of heavy-tailed distributions; it fails for any light-tailed distribution. Indeed, consider any non-negative random variable ξ having a light-tailed distribution, i.e., $\mathbb{E}e^{\lambda \xi} < \infty$ for some $\lambda > 0$. Take $g(x) = \ln x$. If $h(x) = o(x)$ as $x \to \infty$, then $h(x) \leq c + \lambda x/2$ for some $c < \infty$ and, hence,

$$\mathbb{E}e^{h(\xi)+g(\xi)} \leq \mathbb{E}\xi e^{c+\lambda\xi/2} < \infty.$$

Proof (of Theorem 2.9). We will construct a piecewise linear function $h(x)$. To do so we construct two positive sequences $x_n \uparrow \infty$ and $\varepsilon_n \downarrow 0$ as $n \to \infty$ and let

$$h(x) = h(x_{n-1}) + \varepsilon_n(x - x_{n-1}) \quad \text{if } x \in (x_{n-1}, x_n], \ n \geq 1.$$

This function is monotone, since $\varepsilon_n > 0$. Moreover, this function is concave, due to the monotonicity of ε_n.

Put $x_0 = 0$ and $h(0) = 0$. Since ξ is heavy-tailed and $g(x) \to \infty$, we can choose x_1 sufficiently large that $e^{g(x)} \geq 2$ for all $x > x_1$ and

$$\mathbb{E}\{e^\xi; \xi \in (x_0, x_1]\} + e^{x_1}\overline{F}(x_1) > \overline{F}(x_0) + 1.$$

Choose $\varepsilon_1 > 0$ so that

$$\mathbb{E}\{e^{\varepsilon_1\xi}; \xi \in (x_0, x_1]\} + e^{\varepsilon_1 x_1}\overline{F}(x_1) = \overline{F}(0) + 1/2,$$

which is equivalent to

$$\mathbb{E}\{e^{h(\xi)}; \xi \in (x_0, x_1]\} + e^{h(x_1)}\overline{F}(x_1) = e^{h(x_0)}\overline{F}(0) + 1/2.$$

By induction we construct an increasing sequence x_n and a decreasing sequence $\varepsilon_n > 0$ such that $e^{g(x)} \geq 2^n$ for all $x > x_n$ and

$$\mathbb{E}\{e^{h(\xi)}; \xi \in (x_{n-1}, x_n]\} + e^{h(x_n)}\overline{F}(x_n) = e^{h(x_{n-1})}\overline{F}(x_{n-1}) + 1/2^n$$

for any $n \geq 2$. For $n = 1$ this is already done. Make the induction hypothesis for some $n \geq 2$. Due to the heavy-tailedness of ξ and to the convergence $g(x) \to \infty$, there exists x_{n+1} so large that $e^{g(x)} \geq 2^{n+1}$ for all $x > x_{n+1}$ and

$$\mathbb{E}\{e^{\varepsilon_n(\xi - x_n)}; \xi \in (x_n, x_{n+1}]\} + e^{\varepsilon_n(x_{n+1} - x_n)}\overline{F}(x_{n+1}) > 2.$$

As a function of ε_{n+1}, the sum

$$\mathbb{E}\{e^{\varepsilon_{n+1}(\xi - x_n)}; \xi \in (x_n, x_{n+1}]\} + e^{\varepsilon_{n+1}(x_{n+1} - x_n)}\overline{F}(x_{n+1})$$

is continuously decreasing to $\overline{F}(x_n)$ as $\varepsilon_{n+1} \downarrow 0$. Therefore, we can choose $\varepsilon_{n+1} \in (0, \varepsilon_n)$ so that

$$\mathbb{E}\{e^{\varepsilon_{n+1}(\xi - x_n)}; \xi \in (x_n, x_{n+1}]\} + e^{\varepsilon_{n+1}(x_{n+1} - x_n)}\overline{F}(x_{n+1}) = \overline{F}(x_n) + 1/(2^{n+1}e^{h(x_n)}).$$

By the definition of $h(x)$ this is equivalent to the following equality:

$$\mathbb{E}\{e^{h(\xi)}; \xi \in (x_n, x_{n+1}]\} + e^{h(x_{n+1})}\overline{F}(x_{n+1}) = e^{h(x_n)}\overline{F}(x_n) + 1/2^{n+1}.$$

Our induction hypothesis now holds with $n + 1$ in place of n as required.

Next, for any N,

$$\mathbb{E}\{e^{h(\xi)}; \xi \leq x_{N+1}\} = \sum_{n=0}^{N} \mathbb{E}\{e^{h(\xi)}; \xi \in (x_n, x_{n+1}]\}$$

$$= \sum_{n=0}^{N} \left(e^{h(x_n)}\overline{F}(x_n) - e^{h(x_{n+1})}\overline{F}(x_{n+1}) + 1/2^{n+1}\right)$$

$$\leq e^{h(x_0)}\overline{F}(x_0) + 1.$$

Hence, $\mathbb{E}e^{h(\xi)}$ is finite. On the other hand, since $e^{g(x)} \geq 2^n$ for all $x > x_n$,

$$\mathbb{E}\{e^{h(\xi) + g(\xi)}; \xi > x_n\} \geq 2^n \mathbb{E}\{e^{h(\xi)}; \xi > x_n\}$$

$$\geq 2^n \left(\mathbb{E}\{e^{h(\xi)}; \xi \in (x_n, x_{n+1}]\} + e^{h(x_{n+1})}\overline{F}(x_{n+1})\right)$$

$$= 2^n \left(e^{h(x_n)}\overline{F}(x_n) + 1/2^{n+1}\right).$$

Then $\mathbb{E}\{e^{h(\xi) + g(\xi)}; \xi > x_n\} \geq 1/2$ for any n, which implies $\mathbb{E}e^{h(\xi) + g(\xi)} = \infty$. Note also that necessarily $\lim_{n \to \infty} \varepsilon_n = 0$; otherwise $\liminf_{x \to \infty} h(x)/x > 0$ and ξ is light tailed.
□

The latter theorem can be strengthened in the following way (for a proof see [20]):

Theorem 2.10. *Let $\xi \geq 0$ be a random variable with a heavy-tailed distribution. Let $f : \mathbb{R}^+ \to \mathbb{R}$ be a concave function such that $\mathbb{E}e^{f(\xi)} = \infty$. Let the function $g : \mathbb{R}^+ \to \mathbb{R}$ be such that $g(x) \to \infty$ as $x \to \infty$. Then there exists a concave function $h : \mathbb{R}^+ \to \mathbb{R}^+$ such that $h \leq f$, $\mathbb{E}e^{h(\xi)} < \infty$, and $\mathbb{E}e^{h(\xi) + g(\xi)} = \infty$.*

2.3 Lower Limit for Tails of Convolutions

Recall that the *convolution* $F * G$ of any two distributions F and G is given by, for any Borel set B,

$$(F * G)(B) = \int_{-\infty}^{\infty} F(B - y)G(dy) = \int_{-\infty}^{\infty} G(B - y)F(dy),$$

where $B - y = \{x - y : x \in B\}$. If, on some probability space with probability measure $\mathbb{P}$, ξ and η are independent random variables with respective distributions F and G, then $(F * G)(B) = \mathbb{P}\{\xi + \eta \in B\}$. The tail function of the convolution, the *convolution tail*, of F and G is then given by, for any $x \in \mathbb{R}$,

$$\overline{F * G}(x) = \mathbb{P}\{\xi + \eta > x\} = \int_{-\infty}^{\infty} \overline{F}(x - y)G(dy) = \int_{-\infty}^{\infty} \overline{G}(x - y)F(dy).$$

Now let F be a distribution on $\mathbb{R}^+$. In this section we discuss the following lower limit:

$$\liminf_{x \to \infty} \frac{\overline{F * F}(x)}{\overline{F}(x)},$$

in the case where F is heavy-tailed. We start with the following result, which generalises an observation in the Introduction.

Theorem 2.11. *Let* $F_1, \ldots, F_n$ *be distributions on* $\mathbb{R}^+$ *with unbounded supports. Then*

$$\liminf_{x \to \infty} \frac{\overline{F_1 * \ldots * F_n}(x)}{\overline{F}_1(x) + \ldots + \overline{F}_n(x)} \geq 1.$$

Proof. Let $\xi_1, \ldots, \xi_n$ be independent random variables with respective distributions $F_1, \ldots, F_n$. Since the events $\{\xi_k > x, \xi_j \in [0, x] \text{ for all } j \neq k\}$ are disjoint for different k, the convolution tail can be bounded from below in the following way:

$$\overline{F_1 * \ldots * F_n}(x) \geq \sum_{k=1}^{n} \mathbb{P}\{\xi_k > x, \xi_j \in [0, x] \text{ for all } j \neq k\}$$

$$= \sum_{k=1}^{n} \overline{F}_k(x) \prod_{j \neq k} F_j(x)$$

$$\sim \sum_{k=1}^{n} \overline{F}_k(x) \quad \text{as } x \to \infty,$$

which implies the desired statement.   □

Note that in the above proof we have heavily used the condition $F_k(\mathbb{R}^+) = 1$; for distributions on the whole real line $\mathbb{R}$ Theorem 2.11 in general fails.

It follows in particular that, for any distribution F on $\mathbb{R}^+$ with unbounded support and for any $n \geq 2$,

$$\liminf_{x \to \infty} \frac{\overline{F^{*n}}(x)}{\overline{F}(x)} \geq n. \qquad (2.6)$$

In particular,

$$\liminf_{x \to \infty} \frac{\overline{F * F}(x)}{\overline{F}(x)} \geq 2. \qquad (2.7)$$

As already discussed in the Introduction, in the light-tailed case the limit given by the left side of (2.7) is typically greater than 2. For example, for an exponential distribution it equals infinity. Thus we may ask under what conditions do we have equality in (2.7). We show that heavy-tailedness of F is sufficient.

Theorem 2.12. *Let F be a heavy-tailed distribution on $\mathbb{R}^+$. Then*

$$\liminf_{x \to \infty} \frac{\overline{F * F}(x)}{\overline{F}(x)} = 2. \qquad (2.8)$$

Proof. By the lower bound (2.7), it remains to prove the upper bound only, i.e.,

$$\liminf_{x \to \infty} \frac{\overline{F * F}(x)}{\overline{F}(x)} \leq 2.$$

Assume the contrary, i.e. there exist $\delta > 0$ and x_0 such that

$$\overline{F * F}(x) \geq (2 + \delta)\overline{F}(x) \quad \text{for all } x > x_0. \qquad (2.9)$$

Applying Theorem 2.9 with $g(x) = \ln x$, we can choose an increasing concave function $h : \mathbb{R}^+ \to \mathbb{R}^+$ such that $\mathbb{E}e^{h(\xi)} < \infty$ and $\mathbb{E}\xi e^{h(\xi)} = \infty$. For any positive $b > 0$, consider the concave function

$$h_b(x) := \min(h(x), bx).$$

Since F is heavy-tailed, $h(x) = o(x)$ as $x \to \infty$; therefore, for any fixed b there exists x_1 such that $h_b(x) = h(x)$ for all $x > x_1$. Hence, $\mathbb{E}e^{h_b(\xi)} < \infty$ and $\mathbb{E}\xi e^{h_b(\xi)} = \infty$.

For any x, we have the convergence $h_b(x) \downarrow 0$ as $b \downarrow 0$. Then $\mathbb{E}e^{h_b(\xi_1)} \downarrow 1$ as $b \downarrow 0$. Thus there exists b such that

$$\mathbb{E}e^{h_b(\xi_1)} \leq 1 + \delta/4. \qquad (2.10)$$

For any real a and t, put $a^{[t]} = \min(a, t)$. Then

$$\mathbb{E}(\xi_1^{[t]} + \xi_2^{[t]})e^{h_b(\xi_1 + \xi_2)} = 2\mathbb{E}\xi_1^{[t]}e^{h_b(\xi_1 + \xi_2)} \leq 2\mathbb{E}\xi_1^{[t]}e^{h_b(\xi_1) + h_b(\xi_2)},$$

by the concavity of the function h_b. Hence,

$$\frac{\mathbb{E}(\xi_1^{[t]} + \xi_2^{[t]})e^{h_b(\xi_1+\xi_2)}}{\mathbb{E}\xi_1^{[t]}e^{h_b(\xi_1)}} \le 2\frac{\mathbb{E}\xi_1^{[t]}e^{h_b(\xi_1)}\mathbb{E}e^{h_b(\xi_2)}}{\mathbb{E}\xi_1^{[t]}e^{h_b(\xi_1)}}$$

$$= 2\mathbb{E}e^{h_b(\xi_2)} \le 2 + \delta/2, \qquad (2.11)$$

by (2.10). On the other hand, since $(\xi_1 + \xi_2)^{[t]} \le \xi_1^{[t]} + \xi_2^{[t]}$,

$$\frac{\mathbb{E}(\xi_1^{[t]} + \xi_2^{[t]})e^{h_b(\xi_1+\xi_2)}}{\mathbb{E}\xi_1^{[t]}e^{h_b(\xi_1)}} \ge \frac{\mathbb{E}(\xi_1 + \xi_2)^{[t]}e^{h_b(\xi_1+\xi_2)}}{\mathbb{E}\xi_1^{[t]}e^{h_b(\xi_1)}}$$

$$= \frac{\int_0^\infty x^{[t]}e^{h_b(x)}(F*F)(dx)}{\int_0^\infty x^{[t]}e^{h_b(x)}F(dx)}. \qquad (2.12)$$

The right side, after integration by parts, is equal to

$$\frac{\int_0^\infty \overline{F*F}(x)d(x^{[t]}e^{h_b(x)})}{\int_0^\infty \overline{F}(x)d(x^{[t]}e^{h_b(x)})}.$$

Since $\mathbb{E}\xi_1 e^{h_b(\xi_1)} = \infty$, in the latter fraction both the integrals in the numerator and the denominator tend to infinity as $t \to \infty$. For the *increasing* function $h_b(x)$, together with the assumption (2.9) this implies that

$$\liminf_{t\to\infty} \frac{\int_0^\infty \overline{F*F}(x)d(x^{[t]}e^{h_b(x)})}{\int_0^\infty \overline{F}(x)d(x^{[t]}e^{h_b(x)})} \ge 2 + \delta.$$

Substituting this into (2.12) we get a contradiction to (2.11) for sufficiently large t. $\square$

It turns out that the "lim inf" given by the left side of (2.7) is equal to 2 not only for heavy-tailed but also for some light-tailed, distributions. Here is an example. Let F be an atomic distribution at the points x_n, $n = 0, 1, \ldots$, with masses p_n, i.e., $F\{x_n\} = p_n$. Suppose that $x_0 = 1$ and that $x_{n+1} > 2x_n$ for every n. Then the tail of the convolution $F*F$ at the point $x_n - 1$ is equal to

$$\overline{F*F}(x_n - 1) = (F\times F)([x_n, \infty) \times \mathbb{R}^+) + (F\times F)([0, x_{n-1}] \times [x_n, \infty))$$
$$\sim 2\overline{F}(x_n - 1) \quad \text{as } n \to \infty.$$

Hence,

$$\lim_{n\to\infty} \frac{\overline{F*F}(x_n - 1)}{\overline{F}(x_n - 1)} = 2.$$

From this equality and from (2.7),

$$\liminf_{x\to\infty} \frac{\overline{F*F}(x)}{\overline{F}(x)} = 2. \qquad (2.13)$$

Take now $x_n = 3^n$, $n = 0, 1, \ldots$, and $p_n = ce^{-3^n}$, where c is the normalising constant. Then F is a light-tailed distribution satisfying the relation (2.13).

We conclude this section with the following result for convolutions of non-identical distributions.

Theorem 2.13. *Let F_1 and F_2 be two distributions on $\mathbb{R}^+$ and let the distribution F_1 be heavy-tailed. Then*

$$\liminf_{x \to \infty} \frac{\overline{F_1 * F_2}(x)}{\overline{F}_1(x) + \overline{F}_2(x)} = 1. \tag{2.14}$$

Proof. By Theorem 2.11, the left side of (2.14) is at least 1. Assume now that it is strictly greater than 1. Then there exists $\varepsilon > 0$ such that, for all sufficiently large x,

$$\frac{\overline{F_1 * F_2}(x)}{\overline{F}_1(x) + \overline{F}_2(x)} \geq 1 + 2\varepsilon. \tag{2.15}$$

Consider the distribution $G = (F_1 + F_2)/2$. This distribution is heavy-tailed. By Theorem 2.12 we get

$$\liminf_{x \to \infty} \frac{\overline{G * G}(x)}{\overline{G}(x)} = 2. \tag{2.16}$$

On the other hand, (2.15) and Theorem 2.11 imply that, for all sufficiently large x,

$$
\begin{aligned}
\overline{G * G}(x) &= \frac{\overline{F_1 * F_1}(x) + \overline{F_2 * F_2}(x) + 2\overline{F_1 * F_2}(x)}{4} \\
&\geq \frac{2(1-\varepsilon)\overline{F}_1(x) + 2(1-\varepsilon)\overline{F}_2(x) + 2(1+2\varepsilon)(\overline{F}_1(x) + \overline{F}_2(x))}{4} \\
&= 2(1+\varepsilon/2)\overline{G}(x),
\end{aligned}
$$

which contradicts (2.16). □

2.4 Long-Tailed Functions and Their Properties

Our plan is to introduce and to study the subclass of heavy-tailed distributions which are *long-tailed*. Later on we will study also long-tailedness properties of other characteristics of distributions. Therefore, we find it reasonable to start with a discussion of some generic properties of long-tailed functions.

Definition 2.14. An ultimately positive function f is *long-tailed* if and only if

$$\lim_{x \to \infty} \frac{f(x+y)}{f(x)} = 1, \quad \text{for all } y > 0. \tag{2.17}$$

Clearly if f is long-tailed, then we may also replace y by $-y$ in (2.17).

The following result makes a useful connection.

Lemma 2.15. *The function f is long-tailed if and only if $g(x) := f(\log x)$ (defined for positive x) is slowly varying at infinity, i.e., for any fixed $a > 0$,*

$$\frac{g(ax)}{g(x)} \to 1 \quad as \ x \to \infty.$$

Proof. The proof is immediate from the definition of g since

$$\frac{g(ax)}{g(x)} = \frac{f(\log x + \log a)}{f(\log x)}. \qquad \qquad \square$$

If f is long-tailed, then we also have uniform convergence in (2.17) over y in compact intervals. This is obvious for monotone functions, but in the general case the result follows from the Uniform Convergence Theorem for functions slowly varying at infinity, see Theorem 1.2.1 in [9]. Thus, for any $a > 0$, we have

$$\sup_{|y| \leq a} |f(x) - f(x+y)| = o(f(x)) \quad as \ x \to \infty. \qquad (2.18)$$

We give some quite basic closure properties for the class of long-tailed functions. We shall make frequent use of these—usually without further comment.

Lemma 2.16. *Suppose that the functions $f_1, \ldots, f_n$ are all long-tailed. Then*

(i) *For constants c_1 and c_2 where $c_2 > 0$, the function $f_1(c_1 + c_2 x)$ is long-tailed.*
(ii) *If $f \sim \sum_{k=1}^{n} c_k f_k$ where $c_1, \ldots, c_n > 0$, then f is long-tailed.*
(iii) *The product function $f_1 \cdots f_n$ is long-tailed.*
(iv) *The function $\min(f_1, \ldots, f_n)$ is long-tailed.*
(v) *The function $\max(f_1, \ldots, f_n)$ is long-tailed.*

Proof. The proofs of (i)–(iii) are routine from the definition of long-tailedness.
   For (iv) observe that, for any $a > 0$ and any x, we have

$$\min\left(\frac{f_1(x+a)}{f_1(x)}, \frac{f_2(x+a)}{f_2(x)}\right) \leq \frac{\min\left(f_1(x+a), f_2(x+a)\right)}{\min\left(f_1(x), f_2(x)\right)}$$

$$\leq \max\left(\frac{f_1(x+a)}{f_1(x)}, \frac{f_2(x+a)}{f_2(x)}\right).$$

Since f_1, f_2 are long-tailed the result now follows for the case $n = 2$. The result for general n follows by induction.
   For (v) observe that, analogously to the argument for (iv) above, for any $a > 0$ and any x, we have

$$\min\left(\frac{f_1(x+a)}{f_1(x)}, \frac{f_2(x+a)}{f_2(x)}\right) \leq \frac{\max\left(f_1(x+a), f_2(x+a)\right)}{\max\left(f_1(x), f_2(x)\right)}$$

$$\leq \max\left(\frac{f_1(x+a)}{f_1(x)}, \frac{f_2(x+a)}{f_2(x)}\right),$$

and the result now follows as before. $\square$

We now have the following result.

Lemma 2.17. *Let f be a long-tailed function. Then f is heavy-tailed and, moreover, satisfies the following relation: for every $\lambda > 0$,*

$$\lim_{x \to \infty} f(x)e^{\lambda x} = \infty.$$

Proof. Fix $\lambda > 0$. Since f is long-tailed, $f(x+y) \sim f(x)$ as $x \to \infty$ uniformly in $y \in [0,1]$. Hence, there exists x_0 such that, for all $x \geq x_0$ and $y \in [0,1]$,

$$f(x+y) \geq f(x)e^{-\lambda/2}.$$

Then $f(x_0 + n + y) \geq f(x_0)e^{-\lambda(n+1)/2}$ for all $n \geq 1$ and $y \in [0,1]$, and, therefore,

$$\liminf_{x \to \infty} f(x)e^{\lambda x} \geq f(x_0) \lim_{n \to \infty} e^{-\lambda(n+1)/2}e^{\lambda n} = \infty,$$

so that the lemma now follows. $\qquad\square$

However, it is not difficult to construct a heavy-tailed function f which fails to be sufficiently smooth so as to be long-tailed. Put

$$f(x) = \sum_{n=1}^{\infty} 2^{-n}\mathbb{I}\{2^{n-1} < x \leq 2^n\}.$$

Then, for any $\lambda > 0$,

$$\limsup_{x \to \infty} f(x)e^{\lambda x} \geq \limsup_{n \to \infty} 2^{-n}e^{\lambda 2^n} = \infty,$$

so that f is heavy-tailed. On the other hand,

$$\liminf_{x \to \infty} \frac{f(x+1)}{f(x)} \leq \liminf_{n \to \infty} \frac{f(2^n+1)}{f(2^n)} = \frac{1}{2},$$

which shows that f is not long-tailed.

h-Insensitivity

We now introduce a very important concept of which we shall make frequent subsequent use.

Definition 2.18. Given a strictly positive non-decreasing function h, an ultimately positive function f is called *h-insensitive* (or *h-flat*) if

$$\sup_{|y| \leq h(x)} |f(x+y) - f(x)| = o(f(x)) \quad \text{as } x \to \infty, \text{ uniformly in } |y| \leq h(x). \quad (2.19)$$

It is clear that the relation (2.19) implies that the function f is long-tailed and conversely that any long-tailed function is h-insensitive for any constant function h. The following lemma gives a strong converse result, which we shall use repeatedly in Sect. 2.7 and subsequently throughout the monograph.

Lemma 2.19. *Suppose that the function f is long-tailed. Then there exists a function h such that $h(x) \to \infty$ as $x \to \infty$ and f is h-insensitive.*

Proof. For any integer $n \geq 1$, by (2.18), we can choose x_n such that

$$\sup_{|y| \leq n} |f(x+y) - f(x)| \leq f(x)/n \quad \text{for all } x > x_n.$$

Without loss of generality we may assume that the sequence $\{x_n\}$ is increasing to infinity. Put $h(x) = n$ for $x \in [x_n, x_{n+1}]$. Since $x_n \to \infty$ as $n \to \infty$, we have $h(x) \to \infty$ as $x \to \infty$. By the construction we have

$$\sup_{|y| \leq h(x)} |f(x+y) - f(x)| \leq f(x)/n$$

for all $x > x_n$, which completes the proof. $\qquad\qquad\qquad\qquad\qquad\qquad\qquad\square$

One important use of h-insensitivity is the following. The "natural" definition of long-tailedness of a function f is that of h-insensitivity with respect to any constant function $h(x) = a$ for all x and some $a > 0$. The use of this property in this form would then require that both the statements and the proofs of many results would involve a double limiting operation in which first x was allowed to tend to infinity, with the use of the relation (2.18), and following which a was allowed to tend to infinity. The replacement of the constant a by a function h itself increasing to infinity, but sufficiently slowly that the long-tailed function f is h-insensitive, not only enables two limiting operations to be replaced with a single one in proofs, but also permits simpler, cleaner, and more insightful presentations of many results (a typical example is the all-important Lemma 2.34 in Sect. 2.7).

Now observe that if a long-tailed function f is h-insensitive for some function h and if a further positive non-decreasing function $\hat{h}$ is such that $\hat{h}(x) \leq h(x)$ for all x, then (by definition) f is also $\hat{h}$-insensitive. Two trivial, but important (and frequently used), consequences of the combination of this observation with Lemma 2.19 are given by the following proposition.

Proposition 2.20. (i) *Given a finite collection of long-tailed functions $f_1, \ldots, f_n$, we may choose a single function h, increasing to infinity, with respect to which each of the functions f_i is h-insensitive.*
(ii) *Given any long-tailed function f and any positive non-decreasing function $\hat{h}$, we may choose a function h such that $h(x) \leq \hat{h}(x)$ for all x and f is h-insensitive.*

Proof. For (i), note that for each i we may choose a function h_i, increasing to infinity, such that f_i is h_i-insensitive, and then define h by $h(x) = \min_i h_i(x)$.

For (ii), note that we may take $h(x) = \min(\hat{h}(x), \bar{h}(x))$ where $\bar{h}$ is such that f is $\bar{h}$-insensitive. $\qquad\qquad\qquad\qquad\qquad\qquad\qquad\qquad\qquad\qquad\qquad\square$

Finally we note that a further important use of h-insensitivity is the following. For any given positive function h, increasing to infinity, we may consider the class of those distributions whose (necessarily long-tailed) tail functions are h-insensitive. For varying h, this gives a powerful method for the classification of such distributions, which we explore in detail in Sect. 2.8.

2.5 Long-Tailed Distributions

As discussed in the Introduction, all heavy-tailed distributions likely to be encountered in practical applications are sufficiently regular as to be long-tailed, and it is the latter property, as applied to distributions, which we study in this section.

First, for any distribution F on $\mathbb{R}$, recall that we denote by R the hazard function $R(x) := -\ln \overline{F}(x)$. By definition, R is always a non-decreasing function and

$$R(x+1) - R(x) = -\ln \frac{\overline{F}(x+1)}{\overline{F}(x)}.$$

Definition 2.21. A distribution F on $\mathbb{R}$ is called *long-tailed* if $\overline{F}(x) > 0$ for all x and, for any fixed $y > 0$,

$$\overline{F}(x+y) \sim \overline{F}(x) \quad \text{as } x \to \infty. \tag{2.20}$$

That is, the distribution F is long-tailed if and only if its tail function $\overline{F}$ is a long-tailed function. Note that in (2.20) we may again replace y by $-y$. Further, for a distribution F to be long-tailed it is sufficient to require (2.20) to hold for any one non-zero value of y. Note also that the convergence in (2.20) is again uniform over y in compact intervals.

We shall write $\mathcal{L}$ for the class of long-tailed distributions on $\mathbb{R}$. Clearly $F \in \mathcal{L}$ is a *tail property* of the distribution F, since it depends only on $\{\overline{F}(x) : x \geq x_0\}$ for any finite x_0. Further, it follows from Lemma 2.17 that if the distribution F is long-tailed ($F \in \mathcal{L}$) then $\overline{F}$ is a heavy-tailed function, and so, by Theorem 2.6, F is also a heavy-tailed distribution. However, as the example following Lemma 2.17 shows, a heavy-tailed distribution need not be long-tailed.

The following lemma gives some readily verified equivalent characterisations of long-tailedness.

Lemma 2.22. *Let F be a distribution on $\mathbb{R}$ with right-unbounded support, and let ξ be a random variable with distribution F. Then the following are equivalent:*
  (i) *The distribution F is long-tailed ($F \in \mathcal{L}$).*
 (ii) *For any fixed $y > 0$, $F(x, x+y] = o(\overline{F}(x))$ as $x \to \infty$.*
(iii) *For any fixed $y > 0$, $\mathbb{P}\{\xi > x+y \,|\, \xi > x\} \to 1$ as $x \to \infty$.*
(iv) *The hazard function $R(x)$ satisfies $R(x+1) - R(x) \to 0$ as $x \to \infty$.*

Analogously to Lemma 2.16 we further have the following result.

Lemma 2.23. *Suppose that the distributions $F_1, \ldots, F_n$ are all long-tailed (i.e. belong to the class $\mathcal{L}$) and that $\xi_1, \ldots, \xi_n$ are independent random variables with distributions $F_1, \ldots, F_n$, respectively. Then*

(i) *For any constants c_1 and $c_2 > 0$, the distribution of $c_2\xi_1 + c_1$ is long-tailed.*

(ii) *If $\overline{F}(x) \sim \sum_{k=1}^{n} c_k \overline{F}_k(x)$ where $c_1, \ldots, c_n > 0$, then F is long-tailed.*

(iii) *If $F(x) = \min(F_1(x), \ldots, F_n(x))$, then F is long-tailed.*

(iv) *If $F(x) = \max(F_1(x), \ldots, F_n(x))$, then F is long-tailed.*

(v) *The distribution of $\min(\xi_1, \ldots, \xi_n)$ is long-tailed.*

(vi) *The distribution of $\max(\xi_1, \ldots, \xi_n)$ is long-tailed.*

Proof. The proofs follow from the application of Lemma 2.16 to the corresponding tail functions. In particular (v) and (vi) follow from (i) and (iii) of Lemma 2.16. □

2.6 Long-Tailed Distributions and Integrated Tails

In the study of random walks in particular, a key role is played by the integrated tail distribution, the fundamental properties of which we introduce in this section.

Definition 2.24. For any distribution F on $\mathbb{R}$ such that

$$\int_0^\infty \overline{F}(y)\,dy < \infty, \qquad (2.21)$$

(and hence $\int_x^\infty \overline{F}(y)\,dy < \infty$ for any finite x) we define the *integrated tail distribution* F_I via its tail function by

$$\overline{F}_I(x) = \min\left(1, \int_x^\infty \overline{F}(y)dy\right). \qquad (2.22)$$

Note that if ξ is a random variable with distribution F, then

$$\int_x^\infty \overline{F}(y)dy = \mathbb{E}\{\xi; \xi > x\} - x\mathbb{P}\{\xi > x\} = \mathbb{E}\{\xi - x; \xi > x\}. \qquad (2.23)$$

An associated concept (in renewal theory and in queueing) is the *residual distribution* F_r which is defined for any distribution F on $\mathbb{R}^+$ with finite mean a by:

$$F_r(B) = \frac{1}{a}\int_B \overline{F}(y)dy, \quad B \in \mathcal{B}(\mathbb{R}^+).$$

The integrated tail and residual distributions satisfy the equality $\overline{F}_r(x) = \overline{F}_I(x)/a$ for all sufficiently large x.

The following characterisation will frequently be useful.

Lemma 2.25. *Suppose that the distribution F is such that (2.21) holds. Then F_I is long-tailed if and only if $\overline{F}(x) = o(\overline{F}_I(x))$ as $x \to \infty$.*

Proof. The integrated tail distribution F_I is long-tailed ($F_I \in \mathcal{L}$) if and only if $\overline{F}_I(x) - \overline{F}_I(x+1) = o(\overline{F}_I(x))$, or, equivalently, $\overline{F}_I(x) - \overline{F}_I(x+1) = o(\overline{F}_I(x+1))$. The required result now follows from the inequalities

$$\overline{F}(x+1) \leq \overline{F}_I(x) - \overline{F}_I(x+1) \leq \overline{F}(x),$$

valid for all sufficiently large x. $\square$

Lemma 2.26. *Suppose that the distribution F is long-tailed ($F \in \mathcal{L}$) and such that (2.21) holds. Then F_I is long-tailed as well ($F_I \in \mathcal{L}$) and $\overline{F}(x) = o(\overline{F}_I(x))$ as $x \to \infty$.*

Proof. The long-tailedness of F_I follows from the relations, as $x \to \infty$,

$$\overline{F}_I(x+t) = \int_x^\infty \overline{F}(x+t+y)dy \sim \int_x^\infty \overline{F}(x+y)dy = \overline{F}_I(x),$$

for any fixed t. That $\overline{F}(x) = o(\overline{F}_I(x))$ as $x \to \infty$ now follows from Lemma 2.25. $\square$

The converse assertion, i.e., that long-tailedness of F_I implies long-tailedness of F is not in general true. This is illustrated by the following example.

Example 2.27. Let the distribution F be such that $\overline{F}(x) = 2^{-2n}$ for $x \in [2^n, 2^{n+1})$. Then F is not long-tailed since $\overline{F}(2^n - 1)/\overline{F}(2^n) = 4$ for any n, so that $\overline{F}(x - 1)/\overline{F}(x) \not\to 1$ as $x \to \infty$. But we have $x^{-2} \leq \overline{F}(x) \leq 4x^{-2}$ for any $x > 0$. In particular, $\overline{F}_I(x) \geq x^{-1}$ and thus $\overline{F}(x) = o(\overline{F}_I(x))$ as $x \to \infty$. Thus, by Lemma 2.25, F_I is long-tailed.

We now formulate a more general result which will be needed in the theory of random walks with heavy-tailed increments and is also of some interest in its own right. Let F be a distribution on $\mathbb{R}$ and μ a non-negative measure on $\mathbb{R}^+$ such that

$$\int_0^\infty \overline{F}(t)\mu(dt) < \infty. \tag{2.24}$$

We may then define the distribution F_μ on $\mathbb{R}^+$ given by

$$\overline{F}_\mu(x) := \min\left(1, \int_0^\infty \overline{F}(x+t)\mu(dt)\right), \quad x \geq 0. \tag{2.25}$$

If μ is Lebesgue measure, then F_μ is the integrated tail distribution. We can formulate the same question as for F_I: what type of conditions on F imply long-tailedness of F_μ? The answer is given by the following theorem.

Theorem 2.28. *Let F be a long-tailed distribution. Then F_μ is a long-tailed distribution and, for any fixed $y > 0$,*

$$\overline{F}_\mu(x+y) \sim \overline{F}_\mu(x)$$

as $x \to \infty$ uniformly in all μ satisfying (2.24), i.e.,

$$\inf_{\mu} \inf_{x > x_0} \frac{\overline{F}_\mu(x+y)}{\overline{F}_\mu(x)} \to 1 \quad as\ x_0 \to \infty. \tag{2.26}$$

If, in addition, $\overline{F}(x + h(x)) \sim \overline{F}(x)$ as $x \to \infty$, for some positive function h, then (2.26) holds with $h(x)$ in place of y.

Proof. Fix $\varepsilon > 0$. Since $\overline{F}(x+y+u) \sim \overline{F}(x+u)$ as $x \to \infty$ uniformly in $u \geq 0$, there exists x_0 such that (2.24),

$$\overline{F}(x+y+u) \geq (1-\varepsilon)\overline{F}(x+u) \quad \text{for all } x > x_0.$$

Then, for all $x > x_0$ and μ,

$$\overline{F}_\mu(x+y) = \int_0^\infty \overline{F}(x+y+u)\mu(dy) \geq (1-\varepsilon)\int_0^\infty \overline{F}(x+u)\mu(du) = (1-\varepsilon)\overline{F}_\mu(x).$$

Letting $\varepsilon \to 0$ we obtain the desired result. The same argument holds when y is replaced by $h(x)$. $\qquad\square$

2.7 Convolutions of Long-Tailed Distributions

We know from Theorem 2.11 that for any distributions F and G on the positive half-line $\mathbb{R}^+$

$$\liminf_{x \to \infty} \frac{\overline{F*G}(x)}{\overline{F}(x) + \overline{G}(x)} \geq 1. \tag{2.27}$$

In order to get an analogous result for distributions on the entire real line $\mathbb{R}$, we assume some of those involved to be long-tailed. The assumption of the theorem below seems to be the weakest possible in the absence of conditions on left tails.

Theorem 2.29. *Let the distributions $F_1, \ldots, F_n$ on $\mathbb{R}$ be such that the function $\overline{F}_1(x) + \ldots + \overline{F}_n(x)$ is long-tailed. Then,*

$$\liminf_{x \to \infty} \frac{\overline{F_1 * \ldots * F_n}(x)}{\overline{F}_1(x) + \ldots + \overline{F}_n(x)} \geq 1. \tag{2.28}$$

In particular (2.28) holds whenever each of the distributions F_i is long-tailed.

Proof (cf Theorem 2.11). Let $\xi_1, \ldots, \xi_n$ be independent random variables with respective distributions $F_1, \ldots, F_n$. For any fixed $a > 0$, we have the following lower bound:

$$\overline{F_1 * \ldots * F_n}(x) \geq \sum_{k=1}^{n} \mathbb{P}\{\xi_k > x + (n-1)a, \xi_j \in (-a,x] \text{ for all } j \neq k\}$$

$$= \sum_{k=1}^{n} \overline{F_k}(x + (n-1)a) \prod_{j \neq k} F_j(-a,x]. \tag{2.29}$$

For every $\varepsilon > 0$ there exists a such that $F_j(-a,a] \geq 1 - \varepsilon$ for all j. Thus, for all $x > a$,

$$\overline{F_1 * \ldots * F_n}(x) \geq (1-\varepsilon)^{n-1} \sum_{k=1}^{n} \overline{F_k}(x + (n-1)a).$$

Since the function $\overline{F}_1 + \ldots + \overline{F}_n$ is long-tailed,

$$\liminf_{x \to \infty} \frac{\overline{F_1 * \ldots * F_n}(x)}{\overline{F}_1(x) + \ldots + \overline{F}_n(x)} \geq (1-\varepsilon)^{n-1}.$$

The required result (2.28) now follows by letting $\varepsilon \to 0$. $\square$

For identical distributions, Theorem 2.29 yields the following corollary.

Corollary 2.30. *Let the distribution F on $\mathbb{R}$ be long-tailed ($F \in \mathcal{L}$). Then, for any $n \geq 2$,*

$$\liminf_{x \to \infty} \frac{\overline{F^{*n}}(x)}{\overline{F}(x)} \geq n.$$

We also have the following result for the convolution of a long-tailed distribution F with an arbitrary distribution G, the proof of which is similar in spirit to that of Theorem 2.29.

Theorem 2.31. *Let the distributions F and G on $\mathbb{R}$ be such that F is long-tailed ($F \in \mathcal{L}$). Then,*

$$\liminf_{x \to \infty} \frac{\overline{F * G}(x)}{\overline{F}(x)} \geq 1. \tag{2.30}$$

Proof. Let ξ and η be independent random variables with respective distributions F and G. For any fixed a,

$$\overline{F * G}(x) \geq \mathbb{P}\{\xi > x - a, \eta > a\}$$

$$= \overline{F}(x-a)\overline{G}(a). \tag{2.31}$$

For every $\varepsilon > 0$ there exists a such that $\overline{G}(a) \geq 1 - \varepsilon$. Thus, for all x,

$$\overline{F * G}(x) \geq (1-\varepsilon)\overline{F}(x-a).$$

Since the distribution F is long-tailed, it now follows that

$$\liminf_{x \to \infty} \frac{\overline{F * G}(x)}{\overline{F}(x)} \geq 1 - \varepsilon$$

and the required result (2.30) once more followed by letting $\varepsilon \to 0$. $\square$

We now have the following corollary.

Corollary 2.32. *Let the distribution F on $\mathbb{R}$ be such that F is long-tailed ($F \in \mathcal{L}$) and let the distribution G be such that $\overline{G}(a) = 0$ for some a. Then $\overline{F * G}(x) \sim \overline{F}(x)$ as $x \to \infty$.*

Proof. Since $\overline{G}(a) = 0$, we have $\overline{F * G}(x) \leq \overline{F}(x - a)$. Thus since F is long-tailed we have

$$\limsup_{x \to \infty} \frac{\overline{F * G}(x)}{\overline{F}(x)} \leq 1.$$

Combining this result with the lower bound of Theorem 2.31, we obtain the desired equivalence. $\square$

In order to further study the convolutions of long-tailed distributions, we make repeated use of two fundamental decompositions. Let $h > 0$ and let ξ and η be independent random variables with distributions F and G, respectively. Then the tail function of the convolution of F and G possesses the following decomposition: for $x > 0$,

$$\overline{F * G}(x) = \mathbb{P}\{\xi + \eta > x, \xi \leq h\} + \mathbb{P}\{\xi + \eta > x, \xi > h\}. \tag{2.32}$$

If in addition $h \leq x/2$, then

$$\overline{F * G}(x)$$
$$= \mathbb{P}\{\xi + \eta > x, \xi \leq h\} + \mathbb{P}\{\xi + \eta > x, \eta \leq h\} + \mathbb{P}\{\xi + \eta > x, \xi > h, \eta > h\}, \tag{2.33}$$

since if $\xi \leq h$ and $\eta \leq h$ then $\xi + \eta \leq 2h \leq x$.

Note that

$$\mathbb{P}\{\xi + \eta > x, \xi \leq h\} = \int_{-\infty}^{h} \overline{G}(x - y) F(dy), \tag{2.34}$$

while the probability of the event $\{\xi + \eta > x, \xi > h, \eta > h\}$ is symmetric in F and G, and

$$\mathbb{P}\{\xi + \eta > x, \xi > h, \eta > h\} = \int_{h}^{\infty} \overline{F}(\max(h, x - y)) G(dy)$$
$$= \int_{h}^{\infty} \overline{G}(\max(h, x - y)) F(dy). \tag{2.35}$$

Definition 2.33. Given a strictly positive non-decreasing function h, a distribution F on $\mathbb{R}$ is called *h-insensitive*(or *h-flat*) if its tail function $\overline{F}$ is an h-insensitive function (see Definition 2.18). Since $\overline{F}$ is monotone, this reduces to the requirement that $\overline{F}(x \pm h(x)) \sim \overline{F}(x)$ as $x \to \infty$.

Recall from the results for h-insensitive functions that a distribution F is long-tailed if and only if there exists a function h as above with respect to which F is h-insensitive.

For long-tailed distributions F and G we shall now make particular use of the decomposition (2.33) in which the constant h is replaced by a function h increasing to infinity (with $h(x) < x/2$ for all x) and such that both F and G are h-insensitive.

The following three lemmas are the keys to everything that follows later in this section.

Lemma 2.34. *Suppose that the distribution G on $\mathbb{R}$ is long-tailed ($G \in \mathcal{L}$) and that the positive function h is such that $h(x) \to \infty$ as $x \to \infty$ and G is h-insensitive. Then, for any distribution F, as $x \to \infty$,*

$$\int_{-\infty}^{h(x)} \overline{G}(x-y)F(dy) \sim \overline{G}(x),$$

$$\int_{x-h(x)}^{\infty} \overline{F}(x-y)G(dy) \sim \overline{G}(x).$$

Proof. The existence of the function h is guaranteed by Lemma 2.19. We now have

$$\int_{-\infty}^{h(x)} \overline{G}(x-y)F(dy) \leq \overline{G}(x-h(x)).$$

On the other hand we also have,

$$\int_{-\infty}^{h(x)} \overline{G}(x-y)F(dy) \geq \int_{-h(x)}^{h(x)} \overline{G}(x-y)F(dy)$$
$$\geq F(-h(x), h(x)]\overline{G}(x+h(x))$$
$$\sim \overline{G}(x+h(x)) \quad \text{as } x \to \infty,$$

where the last equivalence follows since $h(x) \to \infty$ as $x \to \infty$. The first result now follows from the choice of the function h. The second result follows similarly: the integral is again bounded from above by $\overline{G}(x-h(x))$ and from below by $\overline{F}(-h(x))\overline{G}(x+h(x))$ and the result follows as previously. $\qquad\square$

Remark 2.35. Note the crucial role played by the monotonicity of the tail function $\overline{G}$ in the proof of Lemma 2.34—something which is not available to us in considering, e.g., densities in Chap. 4.

We now prove a version of Lemma 2.34 which is symmetric in the distributions F and G, and which allows us to get many important results for convolutions—see the further discussion below.

Lemma 2.36. *Suppose that the distributions F and G on $\mathbb{R}$ are such that the sum $\overline{F} + \overline{G}$ of their tail functions is a long-tailed function (equivalently the measure $F + G$ is long-tailed in the obvious sense) and that the positive function h is such that $h(x) \to \infty$ as $x \to \infty$ and $\overline{F} + \overline{G}$ is h-insensitive. Then*

$$\int_{-\infty}^{h(x)} \overline{G}(x-y)F(dy) + \int_{-\infty}^{h(x)} \overline{F}(x-y)G(dy) \sim \overline{G}(x) + \overline{F}(x) \quad \text{as } x \to \infty.$$

Proof. The proof is simply a two-sided version of that for the first assertion of Lemma 2.34. The existence of the function h is again guaranteed by Lemma 2.19. Now note first that, as in the earlier proof,

$$\int_{-\infty}^{h(x)} \overline{G}(x-y)F(dy) + \int_{-\infty}^{h(x)} \overline{F}(x-y)G(dy) \leq \overline{G}(x-h(x)) + \overline{F}(x-h(x)),$$

and second that

$$\int_{-\infty}^{h(x)} \overline{G}(x-y)F(dy) + \int_{-\infty}^{h(x)} \overline{F}(x-y)G(dy)$$

$$\geq \int_{-h(x)}^{h(x)} \overline{G}(x-y)F(dy) + \int_{-h(x)}^{h(x)} \overline{F}(x-y)G(dy)$$

$$\geq F(-h(x), h(x)]\overline{G}(x+h(x)) + G(-h(x), h(x)]\overline{F}(x+h(x))$$

$$\sim \overline{G}(x+h(x)) + \overline{F}(x+h(x)) \quad \text{as } x \to \infty,$$

where the last equivalence follows since $h(x) \to \infty$ as $x \to \infty$. The required result now follows from the choice of the function h. $\qquad\square$

Note that special cases under which $\overline{F} + \overline{G}$ is long-tailed are (a) F and G are both long-tailed—in which case Lemma 2.36 (almost) follows from 2.34, and (b) F is long-tailed and $\overline{G}(x) = o(\overline{F}(x))$ as $x \to \infty$.

In various calculations we need to estimate the "internal" part of the convolution. The following result will be useful.

Lemma 2.37. *Let h be any increasing function on $\mathbb{R}^+$ such that $h(x) \to \infty$. Then, for any distributions F_1, F_2, G_1, and G_2 on $\mathbb{R}$,*

$$\limsup_{x \to \infty} \frac{\mathbb{P}\{\xi_1 + \eta_1 > x, \xi_1 > h(x), \eta_1 > h(x)\}}{\mathbb{P}\{\xi_2 + \eta_2 > x, \xi_2 > h(x), \eta_2 > h(x)\}} \leq \limsup_{x \to \infty} \frac{\overline{F_1}(x)}{\overline{F_2}(x)} \cdot \limsup_{x \to \infty} \frac{\overline{G_1}(x)}{\overline{G_2}(x)},$$

where ξ_1, ξ_2, η_1, and η_2 are independent random variables with respective distributions F_1, F_2, G_1 and G_2.

In particular, in the case where the limits of the ratios $\overline{F_1}(x)/\overline{F_2}(x)$ and $\overline{G_1}(x)/\overline{G_2}(x)$ exist, we have

$$\lim_{x \to \infty} \frac{\mathbb{P}\{\xi_1 + \eta_1 > x, \xi_1 > h(x), \eta_1 > h(x)\}}{\mathbb{P}\{\xi_2 + \eta_2 > x, \xi_2 > h(x), \eta_2 > h(x)\}} = \lim_{x \to \infty} \frac{\overline{F_1}(x)}{\overline{F_2}(x)} \cdot \lim_{x \to \infty} \frac{\overline{G_1}(x)}{\overline{G_2}(x)}.$$

Proof. It follows from (2.35) that

$$\mathbb{P}\{\xi_1 + \eta_1 > x, \xi_1 > h(x), \eta_1 > h(x)\}$$

$$\leq \sup_{z>h(x)} \frac{\overline{F_1}(z)}{\overline{F_2}(z)} \int_{h(x)}^{\infty} \overline{F_2}(\max(h(x), x-y))G_1(dy)$$

$$= \sup_{z>h(x)} \frac{\overline{F_1}(z)}{\overline{F_2}(z)} \int_{h(x)}^{\infty} \overline{G_1}(\max(h(x), x-y))F_2(dy).$$

Similarly,

$$\int_{h(x)}^{\infty} \overline{G_1}(\max(h(x), x-y))F_2(dy)$$

$$\leq \sup_{z>h(x)} \frac{\overline{G_1}(z)}{\overline{G_2}(z)} \int_{h(x)}^{\infty} \overline{G_2}(\max(h(x), x-y))F_2(dy)$$

$$= \sup_{z>h(x)} \frac{\overline{G_1}(z)}{\overline{G_2}(z)} \mathbb{P}\{\xi_2 + \eta_2 > x, \xi_2 > h(x), \eta_2 > h(x)\}.$$

Combining these results and recalling that $h(x) \to \infty$ as $x \to \infty$, we obtain the desired conclusion. □

Definition 2.38. Two distributions F and G with right-unbounded supports are said to be *tail-equivalent* if $\overline{F}(x) \sim \overline{G}(x)$ as $x \to \infty$ (i.e. $\lim_{x\to\infty} \overline{F}(x)/\overline{G}(x) = 1$).

In the next two theorems we provide conditions under which a random shifting preserves tail equivalence.

Theorem 2.39. *Suppose that F_1, F_2 and G are distributions on $\mathbb{R}$ such that $\overline{F_1}(x) \sim \overline{F_2}(x)$ as $x \to \infty$. Suppose further that G is long-tailed. Then $\overline{F_1 * G}(x) \sim \overline{F_2 * G}(x)$ as $x \to \infty$.*

Proof. By Lemma 2.19 we can find a function h such that $h(x) \to \infty$ and

$$\overline{G}(x \pm h(x)) \sim \overline{G}(x) \quad \text{as } x \to \infty,$$

i.e. G is h-insensitive. We use the following decomposition: for $k = 1, 2$,

$$\overline{F_k * G}(x) = \left(\int_{-\infty}^{x-h(x)} + \int_{x-h(x)}^{\infty} \right) \overline{F_k}(x-y)G(dy). \qquad (2.36)$$

It follows from the tail equivalence of F_1 and F_2 that $\overline{F_1}(x-y) \sim \overline{F_2}(x-y)$ as $x \to \infty$ uniformly in $y < x - h(x)$. Thus,

$$\int_{-\infty}^{x-h(x)} \overline{F_1}(x-y)G(dy) \sim \int_{-\infty}^{x-h(x)} \overline{F_2}(x-y)G(dy) \qquad (2.37)$$

as $x \to \infty$. Next, by Lemma 2.34, for $k = 1, 2$,

$$\int_{x-h(x)}^{\infty} \overline{F}_k(x-y)G(dy) \sim \overline{G}(x) \quad \text{as } x \to \infty. \tag{2.38}$$

Substituting (2.37) and (2.38) into (2.36) we obtain the required equivalence $\overline{F_1 * G}(x) \sim \overline{F_2 * G}(x)$. □

Theorem 2.40. *Suppose that F_1, F_2, G_1 and G_2 are distributions on $\mathbb{R}$ such that $\overline{F}_1(x) \sim \overline{F}_2(x)$ and $\overline{G}_1(x) \sim \overline{G}_2(x)$ as $x \to \infty$. Suppose further that the function $\overline{F}_1 + \overline{G}_1$ is long-tailed. Then $\overline{F_1 * G_1}(x) \sim \overline{F_2 * G_2}(x)$ as $x \to \infty$.*

Proof. The conditions of the theorem imply that the function $\overline{F}_2 + \overline{G}_2$ is similarly long-tailed. By Lemma 2.19 and the following remark we can choose a function h such that $h(x) \to \infty$ as $x \to \infty$, $h(x) \le x/2$ and, for $k = 1, 2$,

$$\overline{F}_k(x \pm h(x)) + \overline{G}_k(x \pm h(x)) \sim \overline{F}_k(x) + \overline{G}_k(x) \quad \text{as } x \to \infty,$$

i.e. $\overline{F}_k + \overline{G}_k$ is h-insensitive. We use the following decomposition which follows from (2.33) to (2.35):

$$\overline{F_k * G_k}(x) = \int_{-\infty}^{h(x)} \overline{F}_k(x-y)G_k(dy) + \int_{-\infty}^{h(x)} \overline{G}_k(x-y)F_k(dy)$$
$$+ \int_{h(x)}^{\infty} \overline{F}_k(\max(h(x), x-y))G_k(dy). \tag{2.39}$$

Since F_1 and F_2 are tail equivalent and G_1 and G_2 are tail equivalent, it follows from Lemma 2.37 that, as $x \to \infty$,

$$\int_{h(x)}^{\infty} \overline{F}_1(\max(h(x), x-y))G_1(dy) \sim \int_{h(x)}^{\infty} \overline{F}_2(\max(h(x), x-y))G_2(dy). \tag{2.40}$$

Further, by Lemma 2.36, for $k = 1, 2$ and as $x \to \infty$,

$$\int_{-\infty}^{h(x)} \overline{F}_k(x-y)G_k(dy) + \int_{-\infty}^{h(x)} \overline{G}_k(x-y)F_k(dy) \sim \overline{F}_k(x) + \overline{G}_k(x). \tag{2.41}$$

Substituting (2.40) and (2.41) into (2.39) we obtain the required equivalence $\overline{F_1 * G_1}(x) \sim \overline{F_2 * G_2}(x)$. □

We now use Theorem 2.40 to show that the class $\mathcal{L}$ is closed under convolutions. This is a corollary of the following result.

Theorem 2.41. *Suppose that the distributions F and G are such that F is long-tailed and the measure $F + G$ is also long-tailed (i.e. the sum $\overline{F} + \overline{G}$ of the tail functions of the two distributions is long-tailed). Then the convolution $F * G$ is also long-tailed.*

Proof. Fix $y > 0$. Take $F_1 = F$ and F_2 to be equal to F shifted by $-y$, i.e., $\overline{F}_2(x) = \overline{F}(x+y)$. Then $F_2 * G$ is equal to $F * G$ shifted by $-y$. Since F is long-tailed, $\overline{F}_1(x) \sim \overline{F}_2(x)$. Since also $\overline{F} + \overline{G}$ is long-tailed, it follows from Theorem 2.40 with $G_1 = G_2 = G$ that $\overline{F_1 * G}(x) \sim \overline{F_2 * G}(x)$. Hence $\overline{F * G}(x) \sim \overline{F * G}(x+y)$ as $x \to \infty$. □

Both the following corollaries are now immediate from Theorem 2.41 since in each case the measure $F + G$ is long-tailed.

Corollary 2.42. *Let the distributions F and G be long-tailed. Then the convolution $F * G$ is also long-tailed.*

Corollary 2.43. *Suppose that F and G are distributions and that F is long-tailed. Suppose also that $\overline{G}(x) = o(\overline{F}(x))$ as $x \to \infty$. Then $F * G$ is long-tailed.*

Finally in this section we have the following converse result.

Lemma 2.44. *Let F and G be two distributions on $\mathbb{R}^+$ such that F has unbounded support and G is non-degenerate at 0. Suppose that $\overline{G}(x) \leq c\overline{F}(x)$ for some $c < \infty$ and*

$$\limsup_{x \to \infty} \frac{\overline{F * G}(x)}{\overline{F}(x) + \overline{G}(x)} \leq 1. \tag{2.42}$$

Then F is long-tailed.

Proof. Take any a such that $G(a, \infty) > 0$ which is possible because G is not concentrated at 0. Since for any two distributions on $\mathbb{R}^+$

$$\overline{F * G}(x) = \int_0^x \overline{F}(x - y)G(dy) + \overline{G}(x),$$

it follows from the condition (2.42) that

$$\int_0^x \overline{F}(x - y)G(dy) \leq \overline{F}(x) + o(\overline{F}(x) + \overline{G}(x))$$
$$= \overline{F}(x) + o(\overline{F}(x)) \quad \text{as } x \to \infty,$$

due to the condition $\overline{G}(x) \leq c\overline{F}(x)$. This implies that

$$\int_0^x F(x - y, x]G(dy) = \int_0^x (\overline{F}(x - y) - \overline{F}(x))G(dy)$$
$$= o(\overline{F}(x)) \quad \text{as } x \to \infty.$$

For $x \geq a$, the left side is not less than $F(x - a, x]G(a, x]$, hence $F(x - a, x] = o(\overline{F}(x))$ as $x \to \infty$. The latter relation is equivalent to $\overline{F}(x - a) \sim \overline{F}(x)$ which completes the proof. $\qquad\square$

2.8 *h*-Insensitive Distributions

Let F be a long-tailed distribution ($F \in \mathcal{L}$), i.e. a distribution whose tail function $\overline{F}$ is such that for some (and hence for all) non-zero y, we have $\overline{F}(x + y) \sim \overline{F}(x)$ as $x \to \infty$. We saw in Lemma 2.19 that we can then find a non-decreasing positive function h such that $h(x) \to \infty$ as $x \to \infty$ and

$$\overline{F}(x+y) \sim \overline{F}(x) \quad \text{uniformly in } |y| \le h(x), \tag{2.43}$$

i.e. such that the distribution F is h-insensitive (see Definition 2.33).

In this section we turn this process around: we fix a positive function h which is increasing to infinity and seek to identify those long-tailed distributions which are h-insensitive. By varying the choice of h, we then have an important technique for classifying long-tailed distributions according to the heaviness of their tails and for establishing characteristic properties of various classes of these distributions.

Slowly Varying Distributions

As a first example, consider the function h given by $h(x) = \varepsilon x$ for some $\varepsilon > 0$; then the class of h-insensitive distributions coincides with the class of distributions whose tails are *slowly varying at infinity*, i.e., for any $\varepsilon > 0$,

$$\frac{\overline{F}((1+\varepsilon)x)}{\overline{F}(x)} \to 1 \quad \text{as } x \to \infty. \tag{2.44}$$

These distributions are extremely heavy; in particular they do not possess any finite positive moments, i.e., $\int x^\gamma F(dx) = \infty$ for any $\gamma > 0$. Examples are given by distributions F with the following tail functions:

$$\overline{F}(x) \sim 1/\ln^\gamma x, \ \ \overline{F}(x) \sim 1/(\ln\ln x)^\gamma \ \text{ as } x \to \infty, \quad \gamma > 0.$$

Regularly Varying Distributions

We introduce here the well-known class of regularly varying distributions and consider their insensitivity properties.

Recall that an ultimately positive function f is called *regularly varying at infinity with index* $\alpha \in \mathbb{R}$ if, for any fixed $c > 0$,

$$f(cx) \sim c^\alpha f(x) \quad \text{as } x \to \infty. \tag{2.45}$$

A distribution F on $\mathbb{R}$ is called *regularly varying at infinity with index* $-\alpha < 0$ if $\overline{F}(cx) \sim c^{-\alpha}\overline{F}(x)$ as $x \to \infty$, i.e., $\overline{F}(x)$ is regularly varying at infinity with index $-\alpha < 0$.

Particular examples of regularly varying distributions which were introduced in Sect. 2.1 are the *Pareto*, *Burr* and *Cauchy* distributions.

If a distribution F on $\mathbb{R}^+$ is regularly varying at infinity with index $-\alpha < 0$, then all moments of order $\gamma < \alpha$ are finite, while all moments of order $\gamma > \alpha$ are infinite. The moment of order $\gamma = \alpha$ may be finite or infinite depending on the particular behaviour of the corresponding slowly varying function (see below).

If a function f is regularly varying at infinity with index α, then we have $f(x) = x^\alpha l(x)$ for some slowly varying function l. Hence it follows from the discussion of Sect. 2.4 that, for any positive function h such that $h(x) = o(x)$ as $x \to \infty$, we have $f(x+y) \sim f(x)$ as $x \to \infty$ uniformly in $|y| \leq h(x)$; we shall then say that f is $o(x)$-*insensitive*. Similarly we shall say that a distribution F is $o(x)$-*insensitive* if its tail function $\overline{F}$ is $o(x)$-insensitive. Thus distributions which are regularly varying at infinity are $o(x)$-insensitive.

It turns out that integration preserves regular variation of distribution; this result is formulated next and is known as Karamata's theorem for distribution functions.

Theorem 2.45. *A distribution F is regularly varying with index $-\alpha < -1$ if and only if the integrated tail distribution F_I is regularly varying with index $-\alpha + 1 < 0$. If any holds, then $\overline{F}(x) \sim (\alpha - 1)\overline{F_I}(x)/x$ as $x \to \infty$.*

Proof. Due to monotonicity of $\overline{F}(y)$, for any $c_1 > 1$,

$$\overline{F_I}(x) - \overline{F_I}(c_1 x) = \int_x^{c_1 x} \overline{F}(y)dy \leq (c_1 x - x)\overline{F}(x) \tag{2.46}$$

and, for any $c_2 < 1$,

$$\overline{F_I}(c_2 x) - \overline{F_I}(x) \geq (x - c_2 x)\overline{F}(x). \tag{2.47}$$

Assume that F_I is regularly varying with index $-\alpha + 1$. Then, from (2.46) and by regular variation of F_I, for any $c_1 > 1$,

$$\liminf_{x \to \infty} \frac{x\overline{F}(x)}{\overline{F_I}(x)} \geq \frac{1 - c_1^{1-\alpha}}{c_1 - 1}.$$

Letting $c_1 \downarrow 1$, we get

$$\liminf_{x \to \infty} \frac{x\overline{F}(x)}{\overline{F_I}(x)} \geq \alpha - 1. \tag{2.48}$$

Similarly, from (2.47) we get, for any $c_2 < 1$,

$$\limsup_{x \to \infty} \frac{x\overline{F}(x)}{\overline{F_I}(x)} \leq \frac{c_2^{1-\alpha} - 1}{1 - c_2},$$

Letting $c_2 \uparrow 1$, we get

$$\limsup_{x \to \infty} \frac{x\overline{F}(x)}{\overline{F_I}(x)} \leq \alpha - 1. \tag{2.49}$$

Then (2.48) and (2.49) lead to $x\overline{F}(x) \sim (\alpha - 1)\overline{F_I}(x)$ as $x \to \infty$, which implies regular variation of F with index $-\alpha$.

Assume now that F is a regularly varying distribution with index $-\alpha$. Then, for every $c > 0$,

$$\overline{F_I}(cx) = \int_{cx}^{\infty} \overline{F}(y)dy$$

$$= c^{-1} \int_{x}^{\infty} \overline{F}(z/c)dz$$

$$\sim c^{\alpha-1} \int_{x}^{\infty} \overline{F}(z)dz = c^{\alpha-1}\overline{F_I}(x) \quad \text{as } x \to \infty,$$

which means the regular variation of F_I with index $1 - \alpha$.   □

Intermediate Regularly Varying Distributions

It turns out that the property of $o(x)$-insensitivity characterises a slightly wider class of distributions than that of distributions whose tails are regularly varying, and we now discuss this.

Definition 2.46. A distribution F on $\mathbb{R}$ is called *intermediate regularly varying* if

$$\lim_{\varepsilon \downarrow 0} \liminf_{x \to \infty} \frac{\overline{F}(x(1+\varepsilon))}{\overline{F}(x)} = 1. \tag{2.50}$$

Any regularly varying distribution is intermediate regularly varying. But the latter class is richer. We provide first a simple example. Take any density function g which is regularly varying at infinity with index $-\alpha < -1$. Then, by Karamata's Theorem, the corresponding distribution G will be regularly varying with index $-\alpha + 1 < 0$. Now consider any density function f such that $c_1 g(x) \leq f(x) \leq c_2 g(x)$, for some $0 < c_1 < c_2 < \infty$ and for all x. The corresponding distribution F is intermediate regularly varying because

$$F(x, x(1+\varepsilon)] \leq c_2 G(x, x(1+\varepsilon)] \quad \text{and} \quad \overline{F}(x) \geq c_1 \overline{G}(x).$$

On the other hand, F is not necessarily a regularly varying distribution. We now have the following characterisation result.

Theorem 2.47. *A distribution F on $\mathbb{R}$ is intermediate regularly varying if and only if, for any positive function h such that $h(x) = o(x)$ as $x \to \infty$,*

$$\overline{F}(x+h(x)) \sim \overline{F}(x), \tag{2.51}$$

i.e. if and only if F is $o(x)$-insensitive.

Proof. It is straightforward that if F is intermediate regularly varying, then it is $o(x)$-insensitive. Hence it only remains to prove the reverse implication. Assume, on the contrary, that this implication fails. Thus let F be a distribution which is $o(x)$-insensitive but which fails to be intermediate regularly varying. The function

$$l(\varepsilon) := \liminf_{x \to \infty} \frac{\overline{F}(x(1+\varepsilon))}{\overline{F}(x)}$$

decreases in $\varepsilon > 0$, due to the monotonicity of $\overline{F}$. Therefore, the failure of (2.50) implies that there exists a positive δ such that $l(\varepsilon) \leq 1 - 2\delta$ for any $\varepsilon > 0$. Hence, for any positive integer n, we can find x_n such that

$$\overline{F}(x_n(1+1/n)) \leq (1-\delta)\overline{F}(x_n)$$

Without loss of generality we may assume the sequence $\{x_n\}$ to be increasing. Now put $h(x) = x/n$ for $x \in [x_n, x_{n+1})$. Then $h(x) = o(x)$ as $x \to \infty$. However,

$$\liminf_{x\to\infty} \frac{\overline{F}(x+h(x))}{\overline{F}(x)} \leq \liminf_{n\to\infty} \frac{\overline{F}(x_n+h(x_n))}{\overline{F}(x_n)}$$
$$= \liminf_{n\to\infty} \frac{\overline{F}(x_n(1+1/n))}{\overline{F}(x_n)}$$
$$\leq 1-\delta,$$

which contradicts the $o(x)$-insensitivity of F. $\square$

We now give an attractive probabilistic characterisation of intermediate regularly varying distributions.

Theorem 2.48. *A distribution F on $\mathbb{R}$ is intermediate regularly varying if and only if, for any sequence of independent identically distributed random variables ξ_1, ξ_2, ... with finite positive mean,*

$$\frac{\overline{F}(S_n)}{\overline{F}(n\mathbb{E}\xi_1)} \to 1 \quad as \ n \to \infty \qquad (2.52)$$

with probability 1, where $S_n = \xi_1 + \ldots + \xi_n$.

Proof. We suppose first that F is intermediate regularly varying; let $\xi_1, \xi_2, \ldots$ be any sequence of independent identically distributed random variables with finite positive mean, and, for each n, let $S_n = \xi_1 + \ldots + \xi_n$; we show that then the relation (2.52) holds. Let $a = \mathbb{E}\xi_1$. Fix any $\varepsilon > 0$. It follows from the definition of intermediate regular variation that there is n_0 and a $\delta > 0$ such that

$$\sup_{n \geq n_0} \left| \frac{\overline{F}(n(a \pm \delta))}{\overline{F}(na)} - 1 \right| \leq \varepsilon.$$

By the Strong Law of Large Numbers, with probability 1, there exists a random number N such that $|S_n - na| \leq n\delta$ for all $n \geq N$. Then, for $n \geq \max\{N, n_0\}$,

$$\left| \frac{\overline{F}(S_n)}{\overline{F}(na)} - 1 \right| \leq \sup_{n \geq n_0} \left| \frac{\overline{F}(n(a \pm \delta))}{\overline{F}(na)} - 1 \right| \leq \varepsilon.$$

Since $\varepsilon > 0$ is arbitrary, this implies the convergence (2.52).

We now prove the converse implication. Assume that the distribution F is not intermediate regularly varying. It is sufficient to construct a sequence of independent

identically distributed random variables $\xi_1, \xi_2, \ldots$ with mean 1, such that the relation
(2.52) fails to hold (where again $S_n = \xi_1 + \ldots + \xi_n$). By Theorem 2.47 F fails to be
$o(x)$-insensitive, and so there exists an $\varepsilon > 0$, an increasing sequence n_k and an
increasing function h with $h(x) = o(x)$ such that

$$\overline{F}(n_k + h(n_k)) \le (1 - \varepsilon)\overline{F}(n_k) \quad \text{for all } k. \tag{2.53}$$

Since $h(x)/x \to 0$, we can choose an increasing subsequence n_{k_m} such that

$$\sum_{m=1}^{\infty} \frac{h(n_{k_m})}{n_{k_m}} < \infty. \tag{2.54}$$

Since h is increasing it follows also that $\sum_{m=1}^{\infty} n_{k_m}^{-1} < \infty$, and so we can define a
random variable ξ taking values on $\{1 \pm h(n_{k_m}), m = 1, 2, \ldots\}$ with probabilities

$$\mathbb{P}\{\xi = 1 - h(n_{k_m})\} = \mathbb{P}\{\xi = 1 + h(n_{k_m})\} = c/n_{k_m}$$

(where c is the appropriate normalising constant). It further follows from (2.54)
that the random variable ξ has a finite mean; moreover, this mean equals 1. Define
the sequence of independent random variables $\xi_1, \xi_2, \ldots$ to each have the same
distribution as ξ. We shall show that

$$\liminf_{m \to \infty} \mathbb{P}\{S_{n_{k_m}} \ge n_{k_m} + h(n_{k_m})\} > 0. \tag{2.55}$$

From this and from (2.53), and since also $\overline{F}$ is non-increasing, it will follow that

$$\liminf_{m \to \infty} \mathbb{P}\{\overline{F}(S_{n_{k_m}}) \le (1 - \varepsilon)\overline{F}(n_k)\} > 0,$$

so that (2.52) cannot hold.

To show (2.55), fix m and consider the events

$$A_j = \bigcap_{i \le n_{k_m}, i \ne j} \{\xi_i \ne 1 \pm h(n_{k_m})\}, \qquad j = 1, \ldots n_{k_m}.$$

Then the events $A_j \cap \{\xi_j = 1 + h(n_{k_m})\}$ are disjoint. Therefore,

$$\mathbb{P}\{S_{n_{k_m}} \ge n_{k_m} + h(n_{k_m})\}$$
$$\ge \sum_{j=1}^{n_{k_m}} \mathbb{P}\{S_{n_{k_m}} \ge n_{k_m} + h(n_{k_m}) \,|\, A_j, \xi_j = 1 + h(n_{k_m})\} \mathbb{P}\{A_j, \xi_j = 1 + h(n_{k_m})\}$$
$$= n_{k_m} \mathbb{P}\{S_{n_{k_m}} - n_{k_m} \ge h(n_{k_m}) \,|\, A_1, \xi_1 - 1 = h(n_{k_m})\} \mathbb{P}\{A_1\} \mathbb{P}\{\xi_1 = 1 + h(n_{k_m})\}$$
$$= c \mathbb{P}\{S_{n_{k_m}} - n_{k_m} \ge h(n_{k_m}) \,|\, A_1, \xi_1 - 1 = h(n_{k_m})\} \mathbb{P}\{A_1\},$$

where the final equality follows from the definition of the distribution of ξ_1. Using again the independence of the random variables ξ_i, we have

$$\mathbb{P}\{S_{n_{k_m}} - n_{k_m} \geq h(n_{k_m}) | A_1, \xi_1 - 1 = h(n_{k_m})\}$$
$$= \mathbb{P}\{S_{n_{k_m}} - n_{k_m} - (\xi_1 - 1) \geq 0 | A_1\} \geq 1/2,$$

where the final inequality follows from the symmetry about 1 of the common distribution of the random variables ξ_i. In addition,

$$\mathbb{P}\{A_1\} = \left(\mathbb{P}\{\xi_i \neq 1 \pm h(n_{k_m})\}\right)^{n_{k_m}-1}$$
$$= (1 - 2c/n_{k_m})^{n_{k_m}-1} \to e^{-2c} \quad \text{as } m \to \infty.$$

We thus finally obtain that

$$\liminf_{m \to \infty} \mathbb{P}\{S_{n_{k_m}} \geq n_{k_m} + h(n_{k_m})\} \geq ce^{-2c}/2,$$

so that (2.55) follows. □

Other Heavy-Tailed Distributions

We proceed now to heavy-tailed distributions with thinner tails. For the *lognormal distribution*, one can take $h(x) = o(x/\ln x)$ in order to have *h*-insensitivity. For the *Weibull distribution* with parameter $\alpha \in (0, 1)$, one can take $h(x) = o(x^{1-\alpha})$.

In many practical situations, the class of so-called $\sqrt{x}$-*insensitive distributions*—those which are *h*-insensitive for the function $h(x) = x^{1/2}$—is of special interest. Among these are intermediate regularly varying distributions (in particular regularly varying distributions), lognormal distributions and Weibull distributions with shape parameter $\alpha < 1/2$. The reason for interest in this quite broad class is explained by the following theorem, which should be compared with Theorem 2.48.

Theorem 2.49. *For any distribution F on $\mathbb{R}$, the following assertions are equivalent:*
(i) *F is $\sqrt{x}$-insensitive.*
(ii) *For some (any) sequence of independent identically distributed random variables $\xi_1, \xi_2, \ldots$ with positive mean and with finite positive variance,*

$$\frac{\overline{F}(S_n)}{\overline{F}(n\mathbb{E}\xi_1)} \to 1 \quad as \ n \to \infty \tag{2.56}$$

in probability, where $S_n = \xi_1 + \ldots + \xi_n$.

Proof. To show (i)$\Rightarrow$(ii) suppose that the distribution F is $\sqrt{x}$-insensitive and that the independent identically distributed random variables $\xi_1, \xi_2, \ldots$ have common mean $a > 0$ and finite variance. Fix $\varepsilon > 0$. By the Central Limit Theorem, there exist N and A such that $\mathbb{P}\{|S_n - na| \leq A\sqrt{n}\} \geq 1 - \varepsilon$ for all $n \geq N$. It follows from the definition of $\sqrt{x}$-insensitivity that there is n_0 such that

$$\left| \frac{\overline{F}(na \pm A\sqrt{n}))}{\overline{F}(na)} - 1 \right| \leq \varepsilon \quad \text{for all } n \geq n_0.$$

Then, for $n \geq \max\{N, n_0\}$,

$$\mathbb{P}\left\{ \left| \frac{\overline{F}(S_n)}{\overline{F}(na)} - 1 \right| \leq \varepsilon \right\} \geq \mathbb{P}\{|S_n - na| \leq A\sqrt{n}\} \geq 1 - \varepsilon,$$

which establishes (2.56).

To show (ii)$\Rightarrow$(i) assume that the independent identically distributed random variables $\xi_1, \xi_2, \ldots$ have common mean $a > 0$ and finite variance $\sigma^2 > 0$, but that, on the contrary, the distribution F fails to be $\sqrt{x}$-insensitive. Then there exists $\varepsilon > 0$ and an increasing sequence n_k such that, for all k,

$$\overline{F}(n_k a + \sqrt{n_k \sigma^2}) \leq (1 - \varepsilon)\overline{F}(n_k a).$$

Therefore,

$$\mathbb{P}\left\{ \left| \frac{\overline{F}(S_{n_k})}{\overline{F}(n_k a)} - 1 \right| \geq \varepsilon \right\} \geq \mathbb{P}\{S_{n_k} - n_k a \geq \sqrt{n_k \sigma^2}\} \to \int_1^\infty \frac{e^{-u^2/2}}{\sqrt{2\pi}} du > 0,$$

which contradicts (2.56).   □

We finish this section by observing that the exponential distribution, while itself light-tailed, is, in an obvious sense, on the boundary of the class of such distributions. We may construct examples of long-tailed (and hence heavy-tailed) distributions on $\mathbb{R}^+$, say, whose tails are, in a sense, arbitrarily close to that of the exponential distribution. For example, the distribution with tail function

$$\overline{F}(x) = e^{-cx/\ln^\alpha x}, \ \alpha > 0, \ c > 0,$$

is very close to the exponential distribution but is still long-tailed; indeed one can take the function h of Lemma 2.19 to be any such that $h(x) = o(\ln^\alpha x)$ as $x \to \infty$. Further, if we replace the logarithmic function by the mth iterated logarithm, we obtain again a long-tailed distribution.

2.9 Comments

The lower bound (2.7) may be found in Chistyakov [13] and in Pakes [44].

Theorem 2.12 was proved by Foss and Korshunov in [27]. For earlier results see Rudin [48] and Rogozin [46]. Some generalisations may be found in the papers [19, 20] by Denisov, Foss and Korshunov.

The class of long-tailed distributions (but not the term itself) was introduced by Chistyakov in [13], in the context of branching processes.

Theorem 2.40 generalises a result of Cline [16] where the case $F_1, F_2, G_1, G_2 \in \mathcal{L}$ was considered.

Corollary 2.42 is well known from Embrechts and Goldie [21].

A comprehensive study of the theory of regularly varying functions may be found in Seneta [50] and in Bingham, Goldie and Teugels [9].

2.10 Problems

2.1. Let distribution F on $\mathbb{R}^+$ has a regularly varying tail with index α, i.e., let $\overline{F}(x) = L(x)/x^\alpha$ where function $L(x)$ is slowly varying at infinity. Prove that:

(i) Any power moment of distribution F of order $\gamma < \alpha$ is finite.
(ii) Any power moment of order $\gamma > \alpha$ is infinite.

Show by examples that the moment of order $\gamma = \alpha$ may either exist or not, depending on the tail behaviour of slowly varying function $L(x)$.

2.2. Let distribution F on $\mathbb{R}^+$ have a regularly varying tail with index $\alpha > 0$. Prove the distribution of $\log \xi$ is light-tailed.

2.3. Let $\xi > 0$ be a random variable. Prove that the distribution of $\log \xi$ is light-tailed if and only if ξ has a finite power moment of order α, for some $\alpha > 0$.

2.4. Let distribution F on $\mathbb{R}^+$ have an infinite moment of order $\gamma > 0$. Prove that F is heavy-tailed.

2.5. Let random variable $\xi \geq 0$ be such that $\mathbb{E}e^{\xi^\alpha} = \infty$ for some $\alpha < 1$. Prove that the distribution of ξ is heavy-tailed.

2.6. Let random variable ξ has

(i) exponential;   (ii) normal

distribution. Prove that the distribution of $e^{\alpha\xi}$ is both heavy- and long-tailed, for every $\alpha > 0$.

2.7. *Student's t-distribution.* Assume we do not know the exact formula for its density. By estimating the moments, prove that the distribution of the ratio

$$\frac{\xi}{\sqrt{(\xi_1^2 + \ldots + \xi_n^2)/n}}$$

is heavy-tailed where the independent random variables $\xi, \xi_1, \ldots, \xi_n$ are sampled from the standard normal distribution. Moreover, prove that this distribution is regularly varying at infinity.

Hint: Show that the denominator has a positive density function in the neighbourhood of zero.

2.8. Let $\eta_1, \ldots, \eta_n$ be n positive random variables (we do not assume their independence, in general). Prove that the distribution of $\eta_1 + \ldots + \eta_n$ is heavy-tailed if and only if the distribution of at least one of the summands is heavy-tailed.

2.9. Let $\xi > 0$ and $\eta > 0$ be two random variables with heavy-tailed distributions. Can the minimum $\min(\xi, \eta)$ have a light-tailed distribution?

2.10. Suppose that $\xi_1, \ldots, \xi_n$ are independent random variables with a common distribution F and that

$$\xi_{(1)} \leq \xi_{(2)} \leq \cdots \leq \xi_{(n)}$$

are the order statistics.

(i) For $k \leq n$, prove that the distribution of $\xi_{(k)}$ is heavy-tailed if and only if F is heavy-tailed.
(ii) For $k \leq n-1$, prove that the distribution of $\xi_{(k+1)} - \xi_{(k)}$ is heavy-tailed if and only if F is heavy-tailed.
(iii) Based on (ii) and on Problem 8, prove that $\xi_{(k)} - \xi_{(l)}$ has a heavy-tailed distribution if and only if F is heavy-tailed.

2.11. Let ξ and η be two positive independent random variables. Prove that the distribution of $\xi - \eta$ is heavy-tailed if and only if the distribution of ξ is heavy-tailed.

2.12. Let $\xi_n, n = 1, 2, \ldots$, be independent identically distributed random variables on $\mathbb{R}^+$. Let $\nu \geq 1$ be an independent counting random variable. Let both ξ_1 and ν have light-tailed distributions. Prove that the distribution of random sum $\xi_1 + \xi_2 + \ldots + \xi_\nu$ is light-tailed too.

2.13. Let $\xi_n, n = 1, 2, \ldots$, be independent identically distributed random variables on $\mathbb{R}^+$ such that $\mathbb{P}\{\xi_1 > 0\} > 0$. Let $\nu \geq 1$ be an independent counting random variable. Let ν have heavy-tailed distribution. Prove that the distribution of random sum $\xi_1 + \xi_2 + \ldots + \xi_\nu$ is heavy-tailed.

2.14. Find a light-tailed distribution F such that the distribution of the product $\xi_1 \xi_2$ is heavy-tailed where ξ_1 and ξ_2 are two independent random variables with distribution F.

2.15. Let non-negative random variable ξ has distribution F. Consider a family of distributions $F_x(B) := \mathbb{P}\{\xi \in x + B | \xi > x\}, B \in \mathcal{B}(\mathbb{R}^+)$.

(i) Prove F is long-tailed if and only if $F_x \Rightarrow \infty$, as $x \to \infty$.
(ii) Prove F is h-insensitive if and only if $\xi_x / h(x) \Rightarrow \infty$ as $x \to \infty$ where ξ_x is a random variable with distribution F_x.

2.16. We say that $H(x)$ is a *boundary function* for a long-tailed distribution F if the following condition holds: F is h-insensitive if and only if $h(x) = o(H(x))$ as $x \to \infty$. Find any boundary function for:

(i) A regularly varying distribution with index $\alpha > 0$.
(ii) A standard log-normal distribution.
(iii) A Weibull distribution with tail $\overline{F}(x) = e^{-x^\beta}$ where $0 < \beta < 1$.
(iv) A distribution with tail $\overline{F}(x) = e^{-x/\log(1+x)}, x \geq 0$.

2.17. Prove that a distribution whose tail is slowly varying at infinity does not have a boundary function.

2.18. Let a random variable ξ has the standard normal distribution.

(i) Find all values of $\alpha > 0$ such that the power $|\xi|^\alpha$ has a heavy-tailed distribution.
(ii) Prove the power $|\xi|^\alpha$ has a heavy- and long-tailed distribution for every $\alpha < 0$.

2.19. Let independent random variables $\xi_1, \ldots, \xi_n$ have the standard normal distribution. Find all n such that the product $\xi_1 \cdot \ldots \cdot \xi_n$ is heavy-tailed. For those n, is the product also long-tailed?

2.20. Let independent non-negative random variables $\xi_1, \ldots, \xi_n$ have Weibull distribution with the tail $\overline{F}(x) = e^{-x^\beta}$, $\beta > 0$. Find the values of β, for which the product $\xi_1 \cdot \ldots \cdot \xi_n$ has a heavy-tailed distribution.

2.21. *Perpetuity.* Suppose $\xi_1, \xi_2, \ldots$ are independent identically distributed random variables with common uniform distribution in the interval $[-2, 1]$. Let $S_0 = 0$, $S_n = \xi_1 + \ldots + \xi_n$ and

$$Z = \sum_{n=0}^{\infty} e^{S_n}.$$

(i) Prove Z is finite with probability 1 and that Z has a heavy-tailed distribution.
   Hint: Show that $\mathbb{E}Z^\gamma = \infty$ for some $\gamma > 0$.
(ii) How can the result of (i) be generalised to other distributions of ξ's?

2.22. Let F and G be two distributions on $\mathbb{R}^+$ with finite means a_F and a_G. Prove that, for all sufficiently large x,

$$\overline{(F * G)_I}(x) = \overline{F_I}(x) + \overline{F * G_I}(x) + (a_G - 1)\overline{F}(x)$$
$$= \overline{G_I}(x) + \overline{F_I * G}(x) + (a_F - 1)\overline{G}(x).$$

2.23. Prove the distribution of a random variable $\xi \geq 0$ is $\sqrt{x}$-insensitive if and only if the distribution of $\sqrt{\xi}$ is long-tailed. More generally, prove that the distribution of $\xi \geq 0$ is x^α-insensitive with some $0 < \alpha < 1$ if and only if the distribution of $\xi^{1-\alpha}$ is long-tailed.

2.24. Suppose X_n is a time-homogeneous Markov chain with state space $\mathbb{Z}^+$ and transition probabilities p_{ij}. Let X_n be a skip-free Markov chain, i.e., only the transition probabilities $p_{i,i-1}$, $p_{i,i}$ and $p_{i,i+1}$ are non-zero. Let X_n be positive recurrent with invariant probabilities π_i. Show that

$$\pi_i = \prod_{j=1}^{i} \frac{p_{j-1,j}}{p_{j,j-1}}.$$

Further, assume that $\limsup_{i \to \infty} p_{ii} < 1$.

(i) Prove if $\limsup_{i \to \infty} (p_{i,i+1} - p_{i,i-1}) < 0$, then the invariant distribution is light-tailed.
(ii) Prove if $\limsup_{i \to \infty} (p_{i,i+1} - p_{i,i-1}) = 0$, then the invariant distribution is heavy-tailed.

2.25. Suppose X_n is a time-homogeneous irreducible aperiodic Markov chain with state space $\mathbb{Z}^+$. Let X_n be positive recurrent. Let there exist a state i_0 such that the distribution of the jump from this state is heavy-tailed, i.e.,

$$\mathbb{E}\{e^{\lambda X_1} \mid X_0 = i_0\} = \infty,$$

for every $\lambda > 0$. Prove that the invariant distribution of the Markov chain is also heavy-tailed.

2.26. *Excess process.* Suppose X_n is a time-homogeneous irreducible non-periodic Markov chain with state space $\{1,2,3,\ldots\}$. Let $\mathbb{P}\{X_1 = i-1 \mid X_0 = i\} = 1$ for every $i \geq 2$. Denote by F the distribution of the jump from the state 1, i.e., for every $i \in \mathbb{Z}^+$,

$$F\{i\} = \mathbb{P}\{X_1 = i+1 \mid X_0 = 1\}.$$

(i) Prove this Markov chain is positive recurrent if and only if F has finite mean. Find the corresponding invariant distribution.

(ii) Prove the invariant distribution of this chain is heavy-tailed if and only if F is heavy-tailed.

(iii) Prove the invariant distribution of this chain is long-tailed if and only if the integrated tail distribution F_I is long-tailed.

2.27. Suppose X_n is a time-homogeneous irreducible aperiodic Markov chain with state space $\{1,2,3,\ldots\}$. Let there exist $i_0 \geq 1$ such that $\mathbb{P}\{X_1 = i-1 \mid X_0 = i\} = 1$ for every $i \geq i_0 + 1$. Denote by F_i the distribution of the jump from the state $i, i \in \{1,\ldots,i_0\}$, i.e., for every $j \in \mathbb{Z}^+$,

$$F_i\{j\} = \mathbb{P}\{X_1 = j+i \mid X_0 = i\}.$$

(i) Prove this Markov chain is positive recurrent if and only if all $F_i, i \in \{1,\ldots,i_0\}$, have finite mean. Prove that the invariant probabilities $\{\pi_j\}$ satisfy the equations, for $j \geq i_0 + 1$,

$$\pi_j = \sum_{i=1}^{i_0} \pi_i F_i[j-i,\infty).$$

(ii) Prove the invariant distribution of this chain is heavy-tailed if and only if at least one of $F_i, i \in \{1,\ldots,i_0\}$, is heavy-tailed.

(iii) Prove the invariant distribution of this chain is long-tailed if all the integrated tail distributions $F_{i,I}$ are long-tailed.

Chapter 3
Subexponential Distributions

As we stated in the Introduction, all those heavy-tailed distributions likely to be of use in practical applications are not only long-tailed but possess the additional regularity property of subexponentiality. Essentially this corresponds to good tail behaviour under the operation of convolution. In this chapter, following established tradition, we introduce first subexponential distributions on the positive half-line $\mathbb{R}^+$. It is not immediately obvious from the definition, but it nevertheless turns out, that subexponentiality is a tail property of a distribution. It is thus both natural, and important for many applications, to extend the concept to distributions on the entire real line $\mathbb{R}$. We also study the very useful subclass of subexponential distributions which was originally called $\mathcal{S}^*$ in [32] and which we name *strong subexponential*. In particular this class again contains all those heavy-tailed distributions likely to be encountered in practice.

Different sufficient and necessary conditions for subexponentiality may be found in Sects. 3.5 and 3.6. We also discuss the questions of why not every long-tailed distribution is subexponential and why the subexponentiality of a distribution does not imply subexponentiality of the integrated tail distribution.

In Sect. 3.9 we consider closure properties for the class of subexponential distributions. We conclude with the fundamental uniform upper bound for the tail of the nth convolution of a subexponential distribution known as Kesten's bound.

3.1 Subexponential Distributions on the Positive Half-Line

In the previous chapter we showed in (2.7) that, for any distribution F on $\mathbb{R}^+$ with unbounded support,

$$\liminf_{x \to \infty} \frac{\overline{F * F}(x)}{\overline{F}(x)} \geq 2.$$

S. Foss et al., *An Introduction to Heavy-Tailed and Subexponential Distributions*,
Springer Series in Operations Research and Financial Engineering,
DOI 10.1007/978-1-4614-7101-1_3, © Springer Science+Business Media New York 2013

It was then proved in Theorem 2.12 that, for any heavy-tailed distribution F on $\mathbb{R}^+$,

$$\liminf_{x \to \infty} \frac{\overline{F * F}(x)}{\overline{F}(x)} = 2.$$

In particular, if F is heavy-tailed on $\mathbb{R}^+$ and if

$$\frac{\overline{F * F}(x)}{\overline{F}(x)} \to c \quad \text{as } x \to \infty,$$

where $c \in (0, \infty]$, then necessarily $c = 2$. This observation leads naturally to the following definition.

Definition 3.1. Let F be a distribution on $\mathbb{R}^+$ with unbounded support. We say that F is *subexponential*, and write $F \in \mathcal{S}$, if

$$\overline{F * F}(x) \sim 2\overline{F}(x) \quad \text{as } x \to \infty. \tag{3.1}$$

Now let ξ_1 and ξ_2 be independent random variables on $\mathbb{R}^+$ with common distribution F. Then the above definition is equivalent to stating that F is subexponential if

$$\mathbb{P}\{\xi_1 + \xi_2 > x\} \sim 2\mathbb{P}\{\xi_1 > x\} \quad \text{as } x \to \infty.$$

This last relation may be rewritten as

$$\mathbb{P}\{\xi_1 > x | \xi_1 + \xi_2 > x\} \to 1/2 \quad \text{as } x \to \infty.$$

Further, since we always have the equivalence

$$\mathbb{P}\{\max(\xi_1, \xi_2) > x\} = 1 - (1 - \mathbb{P}\{\xi_1 > x\})^2 \sim 2\mathbb{P}\{\xi_1 > x\}$$

as $x \to \infty$, it follows that F is a subexponential distribution if and only if

$$\mathbb{P}\{\xi_1 + \xi_2 > x\} \sim \mathbb{P}\{\max(\xi_1, \xi_2) > x\} \quad \text{as } x \to \infty.$$

Finally, since ξ_1, ξ_2 are non-negative, the inequality $\max(\xi_1, \xi_2) > x$ implies also that $\xi_1 + \xi_2 > x$, and so the subexponentiality of their distribution is equivalent to the following relation:

$$\mathbb{P}\{\xi_1 + \xi_2 > x, \max(\xi_1, \xi_2) \le x\} = o(\mathbb{P}\{\xi_1 > x\}) \quad \text{as } x \to \infty. \tag{3.2}$$

That is, for large x, the only significant way in which $\xi_1 + \xi_2$ can exceed x is that either ξ_1 or ξ_2 should itself exceed x. This is the well-known "principle of a single big jump" for sums of subexponentially distributed random variables.

Lemma 2.44 with $G = F$ implies immediately the following result.

Lemma 3.2. *Any subexponential distribution on $\mathbb{R}^+$ is long-tailed. In particular, any subexponential distribution is heavy-tailed.*

The converse is not true; there exist some long-tailed distributions on $\mathbb{R}^+$ which are not subexponential; see Sect. 3.7 for more detail.

Since a long-tailed distribution F satisfies $\overline{F}(x)e^{\lambda x} \to \infty$ as $x \to \infty$, for all $\lambda > 0$ (see Lemma 2.17), it is this property that originally suggested the name subexponential. However, the name is now always used in the slightly more restrictive sense that we have defined.

In the class of distributions on the positive half-line, subexponentiality is a *tail property*, as are both heavy- and long-tailedness. To see this, observe that if a distribution F_1 on $\mathbb{R}^+$ is subexponential (and therefore long-tailed) and if a distribution F_2 on $\mathbb{R}^+$ is such that, for some x_0, we have $\overline{F}_1(x) = \overline{F}_2(x)$ for all $x \geq x_0$, then by Theorem 2.40 $\overline{F_1 * F_1}(x) \sim \overline{F_2 * F_2}(x)$ as $x \to \infty$, which implies the subexponentiality of F_2.

3.2 Subexponential Distributions on the Whole Real Line

In the previous section we defined subexponential distributions on the positive half-line $\mathbb{R}^+$. We showed there that subexponentiality was a tail property of a distribution. Thus, as remarked at the beginning of this chapter, it is both natural and desirable to extend the concept to distributions on the entire real line $\mathbb{R}$.

The problem is now that of extending the definition appropriately. It turns out that, for a distribution F on the entire real line $\mathbb{R}$, the condition (3.1) no longer defines a tail property of that distribution, nor even implies that the distribution is long-tailed. This is illustrated by the following example.

Example 3.3. For $A \geq 0$, consider the distribution F on the interval $[-A, \infty)$ with the tail function

$$\overline{F}(x) = (x+A+1)^{-2}e^{-(x+A)}, \quad x \geq -A.$$

The convolution tail is given by

$$\overline{F * F}(x) = \int_{-\infty}^{\infty} \overline{F}(x-y)F(dy)$$

$$= \int_{-\infty}^{x/2} \overline{F}(x-y)F(dy) - \int_{x/2}^{\infty} \overline{F}(x-y)d\overline{F}(y)$$

$$= 2\int_{-\infty}^{x/2} \overline{F}(x-y)F(dy) + (\overline{F}(x/2))^2,$$

after integration by parts. We thus have that, as $x \to \infty$,

$$\overline{F * F}(x) \sim 2e^{-x-A} \int_{-A}^{x/2} (x-y)^{-2}e^{y}F(dy) + o(\overline{F}(x))$$

$$\sim 2x^{-2}e^{-x-A} \int_{-A}^{x/2} e^{y}F(dy) + o(\overline{F}(x))$$

$$\sim 2\overline{F}(x) \int_{-A}^{\infty} e^{y}F(dy).$$

Take A such that $\int_{-A}^{\infty} e^y F(dy) = 1$. Then $\overline{F*F}(x) \sim 2\overline{F}(x)$, but F is not long-tailed and indeed F is light-tailed.

The above example shows that the satisfaction of the condition (3.1) *is not* a tail property for the class of distributions on the whole real line $\mathbb{R}$ – for otherwise the condition would be satisfied by the distribution F^+ (given, as in the Introduction, by $F^+(x) = F(x)$ for $x \geq 0$ and $F^+(x) = 0$ for $x < 0$), and Lemma 3.2 would then guarantee that F^+ was long-tailed in contradiction to the result F is not long-tailed.

Thus the most usual way to define the subexponentiality of a distribution F on the whole real line R is to require that the distribution F^+ on $\mathbb{R}^+$ be subexponential. By Lemma 3.2 and Theorem 2.40 the condition (3.1) then continues to hold – it is simply no longer sufficient for subexponentiality. This approach has the advantage of making it immediately clear that subexponentiality remains a tail property, but the disadvantage of requiring a two-stage definition. We shall see below that an equivalent definition, which we shall make formally, is to require that the distribution F, in additional to satisfying (3.1), is also long-tailed. The asserted equivalence follows from the following lemma.

Lemma 3.4. *Let F be a distribution on $\mathbb{R}$ and let ξ be a random variable with distribution F. Then the following are equivalent:*
  (i) *F is long-tailed and $\overline{F*F}(x) \sim 2\overline{F}(x)$ as $x \to \infty$.*
 (ii) *The distribution F^+ of ξ^+ is subexponential.*
(iii) *The conditional distribution $G(B) := \mathbb{P}\{\xi \in B \,|\, \xi \geq 0\}$ is subexponential.*

Proof. Let ξ_1 and ξ_2 be two independent copies of ξ.

(i)$\Rightarrow$(ii). Suppose that F is long-tailed. Fix $A > 0$. On the event $\{\xi_k > -A\}$, we have $\xi_k^+ \leq \xi_k + A$. Thus, for $x \geq 0$,

$$\mathbb{P}\{\xi_1^+ + \xi_2^+ > x\} \leq \mathbb{P}\{\xi_1 + \xi_2 > x - 2A, \xi_1 > -A, \xi_2 > -A\}$$
$$+ \mathbb{P}\{\xi_2 > x, \xi_1 \leq -A\} + \mathbb{P}\{\xi_1 > x, \xi_2 \leq -A\}$$
$$\leq \mathbb{P}\{\xi_1 + \xi_2 > x - 2A\} + 2\overline{F}(x)F(-A).$$

Hence, since F is long-tailed,

$$\limsup_{x \to \infty} \frac{\mathbb{P}\{\xi_1^+ + \xi_2^+ > x\}}{\overline{F}(x)} \leq \lim_{x \to \infty} \frac{\mathbb{P}\{\xi_1 + \xi_2 > x - 2A\}}{\overline{F}(x - 2A)} + 2F(-A)$$
$$= 2 + 2F(-A).$$

Since A can be chosen as large as we please,

$$\limsup_{x \to \infty} \frac{\mathbb{P}\{\xi_1^+ + \xi_2^+ > x\}}{\overline{F}(x)} \leq 2.$$

Together with (2.7) this implies that $\overline{F^+ * F^+}(x) \sim 2\overline{F^+}(x)$ as $x \to \infty$, i.e. that the distribution F^+ of ξ^+ is subexponential.

(ii)$\Rightarrow$(i). Suppose now that the distribution F^+ of ξ^+ is subexponential. That F^+ and hence F is long-tailed follows from Lemma 3.2. We further have $\xi_1 + \xi_2 \leq \xi_1^+ + \xi_2^+$, so that

$$\overline{F * F}(x) \leq \overline{F^+ * F^+}(x) \sim 2\overline{F}(x)$$

as $x \to \infty$, again by the subexponentiality of F^+. Together with the lower bound for the 'lim inf' provided by Corollary 2.30, we get the required tail asymptotics $\overline{F * F}(x) \sim 2\overline{F}(x)$ as $x \to \infty$.

(ii)$\Leftrightarrow$(iii). We show now the equivalence of the conditions (ii) and (iii). Define first $p = \mathbb{P}\{\xi < 0\}$ and observe that, for $x \geq 0$,

$$
\begin{aligned}
\mathbb{P}\{\xi_1^+ + \xi_2^+ > x\} &= 2\mathbb{P}\{\xi_1 < 0, \xi_2 > x\} + \mathbb{P}\{\xi_1 + \xi_2 > x, \xi_1 \geq 0, \xi_2 \geq 0\} \\
&= 2p\overline{F}(x) + (1-p)^2 \mathbb{P}\{\xi_1 + \xi_2 > x \mid \xi_1 \geq 0, \xi_2 \geq 0\} \\
&= 2p\overline{F}(x) + (1-p)^2 \overline{G * G}(x).
\end{aligned}
$$

Since also, for $x \geq 0$, we have $\overline{F}(x) = (1-p)\overline{G}(x)$, the subexponentiality of G is equivalent to the condition that, as $x \to \infty$,

$$
\begin{aligned}
\mathbb{P}\{\xi_1^+ + \xi_2^+ > x\} &\sim 2p\overline{F}(x) + 2(1-p)\overline{F}(x) \\
&= 2\mathbb{P}\{\xi^+ > x\},
\end{aligned}
$$

i.e. to the subexponentiality of F^+. $\qquad\square$

The above lemma allows us to make the following definition of whole-line subexponentiality.

Definition 3.5. Let F be a distribution on $\mathbb{R}$ with right-unbounded support. We say that F is *whole-line subexponential*, and write $F \in \mathcal{S}_{\mathbb{R}}$, if F is long-tailed and

$$\overline{F * F}(x) \sim 2\overline{F}(x) \quad \text{as } x \to \infty.$$

Equivalently, a random variable ξ has a whole-line subexponential distribution if ξ^+ has a subexponential distribution.

Thus whole-line subexponentiality generalises the concept of subexponentiality on the positive half-line $\mathbb{R}^+$ and any distribution which is subexponential on $\mathbb{R}^+$ or $\mathbb{R}$ is long-tailed, i.e. $\mathcal{S} \subseteq \mathcal{S}_{\mathbb{R}} \subseteq \mathcal{L}$.

We now have the following theorem which provides the foundation for our results on convolutions of subexponential distributions.

Theorem 3.6. *Let the distribution F on $\mathbb{R}$ be long-tailed ($F \in \mathcal{L}$) and let ξ_1, ξ_2 be two independent random variables with distribution F. Let the function h be such that $h(x) \to \infty$ as $\to \infty$ and F is h-insensitive (see Definition 2.33). Then F is whole-line subexponential ($F \in \mathcal{S}_{\mathbb{R}}$) if and only if*

$$\mathbb{P}\{\xi_1 + \xi_2 > x, \xi_1 > h(x), \xi_2 > h(x)\} = o(\overline{F}(x)) \quad \text{as } x \to \infty. \qquad (3.3)$$

Proof. We assume first that additionally $h(x) < x/2$ for all x. Then, for any x,

$$\mathbb{P}\{\xi_1 + \xi_2 > x\}$$
$$= \mathbb{P}\{\xi_1 + \xi_2 > x, \ \xi_1 \le h(x)\} + \mathbb{P}\{\xi_1 + \xi_2 > x, \ \xi_2 \le h(x)\}$$
$$+ \mathbb{P}\{\xi_1 + \xi_2 > x, \ \xi_1 > h(x), \ \xi_2 > h(x)\}. \quad (3.4)$$

Since F is long-tailed, it follows from (2.34), the given conditions on h and Lemma 2.34 that, for $i = 1, 2$,

$$\mathbb{P}\{\xi_1 + \xi_2 > x, \ \xi_i \le h(x)\} \sim \overline{F}(x) \qquad \text{as } x \to \infty. \quad (3.5)$$

Again, since F is long-tailed, the subexponentiality of F is equivalent to the requirement that $\mathbb{P}\{\xi_1 + \xi_2 > x\} \sim 2\overline{F}(x)$ as $x \to \infty$, and the equivalence of this to the condition (3.3) now follows from (3.4) and (3.5).

In the case where we do not have $h(x) < x/2$ for all x, small variations are required to the above proof. If F is subexponential, then we may consider instead the function $\hat{h}$ given by $\hat{h}(x) = \min(h(x), x/2)$. Since F is then also $\hat{h}$-insensitive, the relation (3.3) holds with h replaced by $\hat{h}$, and so also in its original form. Conversely, if (3.3) holds, then that F is subexponential follows as before, except only that we now have "$\le$" instead of equality in (3.4), which does not affect the argument.   □

Theorem 3.6 implies that, as in the case of non-negative subexponential summands, the most probable way for large deviations of the sum $\xi_1 + \xi_2$ to occur is that one summand is small and the other is large; for (very) large x, the main contribution to the probability $\mathbb{P}\{\xi_1 + \xi_2 > x\}$ is made by the probabilities of the events $\{\xi_1 + \xi_2 > x, \ \xi_i \le h(x)\}$ for $i = 1, 2$.

We now give what is almost a restatement of Theorem 3.6 in terms of integrals, in a form which will be of use in various of our subsequent calculations.

Theorem 3.7. *Let the distribution F on $\mathbb{R}$ be long-tailed. Then the following are equivalent:*

(i) *F is whole-line subexponential, i.e. $F \in \mathcal{S}_{\mathbb{R}}$.*

(ii) *For every function h with $h(x) < x/2$ for all x and such that $h(x) \to \infty$ as $x \to \infty$,*

$$\int_{h(x)}^{x-h(x)} \overline{F}(x-y) F(dy) = o(\overline{F}(x)) \ \text{ as } x \to \infty. \quad (3.6)$$

(iii) *There exists a function h with $h(x) < x/2$ for all x, such that $h(x) \to \infty$ as $x \to \infty$ and F is h-insensitive, and the relation (3.6) holds.*

Proof. As remarked above, the theorem is only a slight variation on Theorem 3.6. Let ξ_1 and ξ_2 again be independent random variables with common distribution F, and let h be any function such that $h(x) < x/2$, $h(x) \to \infty$ and F is h-insensitive (note that since $F \in \mathcal{L}$ there is always at least one such function h); the difference between the left side of (3.3) and the left side of (3.6) is

$$\mathbb{P}\{\xi_1 > x - h(x), \ \xi_2 > h(x)\} = \overline{F}(x - h(x))\overline{F}(h(x)) \sim \overline{F}(x)\overline{F}(h(x)) = o(\overline{F}(x))$$

as $x \to \infty$. The theorem thus follows immediately from Theorem 3.6, except only that it is necessary to observe that the reason why, in the statement (ii), we do not require any restriction to functions h such that F is h-insensitive follows from Proposition 2.20(ii). □

In the succeeding sections, we will make use of the following result.

Lemma 3.8. *Suppose that F is whole-line subexponential and that the function h is such that $h(x) \to \infty$ as $x \to \infty$. Let the distributions G_1, G_2 be such that, for $i = 1, 2$, we have $\overline{G_i}(x) = O(\overline{F}(x))$ as $x \to \infty$. If η_1 and η_2 are independent random variables with distributions G_1 and G_2, then*

$$\mathbb{P}\{\eta_1 + \eta_2 > x, \eta_1 > h(x), \eta_2 > h(x)\} = o(\overline{F}(x)) \quad \text{as } x \to \infty.$$

Proof. Let ξ_1 and ξ_2 be two independent random variables with distribution F. Since $\overline{G_i}(x) = O(\overline{F}(x))$, it follows from Lemma 2.37 that, for some $c < \infty$,

$$\mathbb{P}\{\eta_1 + \eta_2 > x, \eta_1 > h(x), \eta_2 > h(x)\} \leq c\mathbb{P}\{\xi_1 + \xi_2 > x, \xi_1 > h(x), \xi_2 > h(x)\}.$$

The subexponentiality of F and Theorem 3.6, together with the immediately preceding remark, imply that

$$\mathbb{P}\{\xi_1 + \xi_2 > x, \xi_1 > h(x), \xi_2 > h(x)\} = o(\overline{F}(x)).$$

Hence the result follows. □

3.3 Subexponentiality and Weak Tail-Equivalence

We start with the definition of weak tail-equivalence and then use this property to establish a number of powerful results.

Definition 3.9. Two distributions F and G with right-unbounded supports are called *weakly tail-equivalent* if there exist $c_1 > 0$ and $c_2 < \infty$ such that, for any $x > 0$,

$$c_1 \leq \frac{\overline{F}(x)}{\overline{G}(x)} \leq c_2.$$

This is equivalent to the condition

$$0 < \liminf_{x \to \infty} \frac{\overline{F}(x)}{\overline{G}(x)} \leq \limsup_{x \to \infty} \frac{\overline{F}(x)}{\overline{G}(x)} < \infty.$$

Lemma 3.10. *Let F and G be weakly tail-equivalent distributions on $\mathbb{R}$. Suppose that either (i) both F and G are long-tailed, or (ii) both F and G are concentrated on $\mathbb{R}^+$, and suppose further that*

$$\limsup_{x \to \infty} \frac{\overline{F * G}(x)}{\overline{F}(x) + \overline{G}(x)} \leq 1. \tag{3.7}$$

Then both F and G are subexponential.

Proof. It follows from Lemma 2.44 that, in both the cases considered, both F and G are long-tailed.

Now let h be any function such that $h(x) < x/2$, $h(x) \to \infty$, and both F and G are h-insensitive (recall that the existence of such a function is guaranteed by the results of Sect. 2.4). Let ξ and η be independent random variables with distributions F and G respectively. It follows from the decomposition (2.33) (with $h(x)$ in place of h), Lemma 2.34, and the condition (3.7) that

$$\mathbb{P}\{\xi + \eta > x, \xi > h(x), \eta > h(x)\} = o(\overline{F}(x) + \overline{G}(x)) \quad \text{as } x \to \infty. \qquad (3.8)$$

Let ξ' be an additional random variable, independent of ξ, with distribution F. Then, from (3.8), the weak tail-equivalence of F and G, and Lemma 2.37,

$$\mathbb{P}\{\xi + \xi' > x, \xi > h(x), \xi' > h(x)\} = o(\overline{F}(x) + \overline{G}(x))$$
$$= o(\overline{F}(x)) \quad \text{as } x \to \infty,$$

where the second line in the above display again follows from the weak tail-equivalence of F and G. Hence, by Theorem 3.6, F is subexponential. $\qquad\square$

Now we prove that the class of subexponential distributions is closed under the weak tail-equivalence relation.

Theorem 3.11. *Suppose that F is whole-line subexponential, i.e. $F \in \mathcal{S}_\mathbb{R}$, that G is long-tailed, and that F and G are weakly tail-equivalent. Then $G \in \mathcal{S}_\mathbb{R}$.*

Proof. Choose a function h such that $h(x) \to \infty$ and G is h-insensitive. Let η_1 and η_2 be independent random variables with distribution G. Then, from Lemma 3.8 and the given weak tail-equivalence,

$$\mathbb{P}\{\eta_1 + \eta_2 > x, \eta_1 > h(x), \eta_2 > h(x)\} = o(\overline{F}(x)) = o(\overline{G}(x)).$$

Hence it follows from Theorem 3.6 that $G \in \mathcal{S}_\mathbb{R}$. $\qquad\square$

Definition 3.12. Two distributions F and G with right-unbounded supports are said to be *proportionally tail-equivalent* if there exists a constant $c > 0$ such that $\overline{F}(x) \sim c\overline{G}(x)$ as $x \to \infty$.

Theorem 3.11 has the following corollary.

Corollary 3.13. *Let the distributions F and G be proportionally tail-equivalent. If $F \in \mathcal{S}_\mathbb{R}$ then $G \in \mathcal{S}_\mathbb{R}$.*

We now turn to convolutions of many distributions.

Theorem 3.14. *Let (a reference distribution) $F \in \mathcal{S}_\mathbb{R}$. Suppose that distributions $G_1, \ldots, G_n$ are such that, for each i, the function $\overline{F} + \overline{G}_i$ is long-tailed and $\overline{G}_i(x) = O(\overline{F}(x))$ as $x \to \infty$. Then*

$$\overline{G_1 * \cdots * G_n}(x) = \overline{G}_1(x) + \ldots + \overline{G}_n(x) + o(\overline{F}(x)) \quad \text{as } x \to \infty.$$

Proof. Note first that it follows from the conditions of the theorem that, for each i and for any constant a,

$$\overline{F}(x+a) + \overline{G_i}(x+a) = \overline{F}(x) + \overline{G_i}(x) + o(\overline{F}(x) + \overline{G_i}(x))$$
$$= \overline{F}(x) + \overline{G_i}(x) + o(\overline{F}(x)).$$

Hence from the representation $F + \sum_{i=1}^{k} G_i = \sum_{i=1}^{k}(F + G_i) - (k-1)F$ and since F is also long-tailed, for each k and for any constant a,

$$\overline{F}(x+a) + \sum_{i=1}^{k} \overline{G_i}(x+a) = \overline{F}(x) + \sum_{i=1}^{k} \overline{G_i}(x) + o(\overline{F}(x)),$$

and so the measure $F + \sum_{i=1}^{k} G_i$ (i.e. the function $\overline{F} + \sum_{i=1}^{k} \overline{G_i}$) is also long-tailed. Note also that for each k we have $\sum_{i=1}^{k} \overline{G_i}(x) = O(\overline{F}(x))$. It now follows that it is sufficient to prove the theorem for case $n = 2$, the general result then following by induction.

By Lemma 2.19 and Proposition 2.20 there exists a function h such that $h(x) \to \infty$, $h(x) \le x/2$, and F, $F + G_1$ and $F + G_2$ are all h-insensitive. It then follows from Lemma 2.34 that, as $x \to \infty$,

$$\int_{-\infty}^{h(x)} \overline{G_1}(x-y)G_2(dy) = \int_{-\infty}^{h(x)} (\overline{G_1}+\overline{F})(x-y)G_2(dy) - \int_{-\infty}^{h(x)} \overline{F}(x-y)G_2(dy)$$
$$= \overline{G_1 + F}(x) - \overline{F}(x) + o(\overline{G_1}(x) + \overline{F}(x))$$
$$= \overline{G_1}(x) + o(\overline{F}(x)), \tag{3.9}$$

and similarly

$$\int_{-\infty}^{h(x)} \overline{G_2}(x-y)G_1(dy) = \overline{G_2}(x) + o(\overline{F}(x)). \tag{3.10}$$

Further, from Lemma 3.8,

$$\int_{h(x)}^{\infty} \overline{G_1}(\max(h(x), x-y)G_2(dy) = o(\overline{F}(x)). \tag{3.11}$$

The required result now follows from the decomposition (2.33) (where ξ and η are independent random variables with distributions G_1 and G_2) and from (3.9) to (3.11). □

Theorem 3.15. *Suppose again that the conditions of Theorem 3.14 hold, and that additionally $G_1 \in \mathcal{L}$ and that G_1 is weakly tail equivalent to F. Then $G_1 * \cdots * G_n \in \mathcal{S}_{\mathbb{R}}$, and additionally $G_1 * \cdots * G_n$ is weakly tail equivalent to F.*

Proof. It follows from Theorem 3.11 that $G_1 \in \mathcal{S}_{\mathbb{R}}$. Further the weak tail equivalence of F and G_1 implies that, for each k, $\overline{G_k}(x) = O(\overline{G_1}(x))$. Hence by Theorem 3.14 with $F = G_1$, the distribution $G_1 * G_2 * \cdots * G_n$ is long-tailed and weakly tail equivalent to G_1 and so also to F. In particular, again by Theorem 3.11, $G_1 * \cdots * G_n \in \mathcal{S}_{\mathbb{R}}$. □

We have the following corollaries of Theorems 3.14 and 3.15.

Corollary 3.16. *Suppose that distributions F and G are such that $F \in \mathcal{S}_{\mathbb{R}}$, that $\overline{F} + \overline{G}$ is long-tailed and that $\overline{G}(x) = O(\overline{F}(x))$ as $x \to \infty$. Then $F * G \in \mathcal{S}_{\mathbb{R}}$ and*

$$\overline{F * G}(x) = \overline{F}(x) + \overline{G}(x) + o(\overline{F}(x)) \quad \text{as } x \to \infty.$$

Proof. This result follows from Theorems 3.14 and 3.15 in the case $n = 2$ with G_1 replaced by F and G_2 by G. $\qquad\square$

Corollary 3.17. *Assume that F, $G \in \mathcal{S}_{\mathbb{R}}$. If F and G are weakly tail-equivalent, then $F * G \in \mathcal{S}_{\mathbb{R}}$.*

Corollary 3.18. *Assume that $F \in \mathcal{S}_{\mathbb{R}}$. If $\overline{G}(x) = o(\overline{F}(x))$ as $x \to \infty$, then $F * G \in \mathcal{S}_{\mathbb{R}}$ and $\overline{F * G}(x) \sim \overline{F}(x)$.*

Corollary 3.19. *Suppose that $F \in \mathcal{S}_{\mathbb{R}}$. Let $G_1, \ldots, G_n$ be distributions such that $\overline{G_i}(x)/\overline{F}(x) \to c_i$ as $x \to \infty$, for some constants $c_i \geq 0$, $i = 1, \ldots, n$. Then*

$$\frac{\overline{G_1 * \ldots * G_n}(x)}{\overline{F}(x)} \to c_1 + \ldots + c_n \quad \text{as } x \to \infty.$$

If $c_1 + \ldots + c_n > 0$, then $G_1 * \ldots * G_n \in \mathcal{S}_{\mathbb{R}}$.

Proof. The first statement of the corollary is immediate from Theorem 3.14. If $c_1 + \ldots + c_n > 0$, we may assume without loss of generality that $c_1 > 0$, so that the second statement follows from Theorem 3.15. $\qquad\square$

The following result is a special case of Corollary 3.19.

Corollary 3.20. *Assume that $F \in \mathcal{S}_{\mathbb{R}}$. Then for any $n \geq 2$, $\overline{F^{*n}}(x)/\overline{F}(x) \to n$ as $x \to \infty$. In particular, $F^{*n} \in \mathcal{S}_{\mathbb{R}}$.*

The following converse result follows.

Theorem 3.21. *Let a distribution F on $\mathbb{R}^+$ with unbounded support be such that $\overline{F^{*n}}(x) \sim n\overline{F}(x)$ for some $n \geq 2$. Then F is subexponential.*

Proof. Take $G := F^{*(n-1)}$. For any x we have the inequality $\overline{G}(x) \geq \overline{F}(x)$. On the other hand, $\overline{G}(x) \leq \overline{F^{*n}}(x) \sim n\overline{F}(x)$. Hence the distributions F and G are weakly tail-equivalent. Thus by Theorem 2.11, as $x \to \infty$,

$$\overline{F * G}(x) \geq (1 + o(1))(\overline{F}(x) + \overline{G}(x))$$
$$= \overline{F}(x) + \overline{G}(x) + o(\overline{F}(x)).$$

Recalling that $\overline{F * G}(x) = \overline{F^{*n}}(x) \sim n\overline{F}(x)$, we deduce the following upper bound:

$$\overline{F^{*(n-1)}}(x) = \overline{G}(x) \leq (n - 1 + o(1))\overline{F}(x).$$

Together with lower bound (2.6) this implies that $\overline{F^{*(n-1)}}(x) \sim (n-1)\overline{F}(x)$ as $x \to \infty$. By induction we deduce then that $\overline{F^{*2}}(x) \sim 2\overline{F}(x)$, which completes the proof. $\qquad\square$

3.4 The Class S^* of Strong Subexponential Distributions

We have already observed that a heavy-tailed distribution F on $\mathbb{R}^+$ is subexponential if and only if it is long-tailed and its tail is sufficiently regular that $\lim_{x\to\infty} \overline{F*F}(x)/\overline{F}(x)$ exists (and that this limit is then equal to 2). Thus subexponentiality, with all its important properties for the tails of convolutions, is effectively guaranteed for all those heavy-tailed distributions likely to be encountered in practice.

However, some applications, for example, those concerned with the behaviour of the maxima of random walks with heavy-tailed increments, require a very slightly stronger regularity condition with respect to their tails—that of membership of the class S^* of strong subexponential distributions which we introduce below. We shall see that membership of S^* is again a tail property of a distribution and that S^* is a subclass of the class $S_{\mathbb{R}}$ of distributions which are whole-line subexponential.

For any distribution F on $\mathbb{R}$ with right-unbounded support, we have the inequality

$$\int_0^x \overline{F}(x-y)\overline{F}(y)dy = 2\int_0^{x/2} \overline{F}(x-y)\overline{F}(y)dy$$

$$\geq 2\overline{F}(x)\int_0^{x/2} \overline{F}(y)dy.$$

Therefore, always

$$\liminf_{x\to\infty} \frac{1}{\overline{F}(x)} \int_0^x \overline{F}(x-y)\overline{F}(y)dy \geq 2m,$$

where $m = \mathbb{E}\xi^+$ and ξ has distribution F. If F is heavy-tailed, then (see Lemma 4 in [27])

$$\liminf_{x\to\infty} \frac{1}{\overline{F}(x)} \int_0^x \overline{F}(x-y)\overline{F}(y)dy = 2m. \tag{3.12}$$

These observations provide a motivation for the following definition.

Definition 3.22. Let F be a distribution on $\mathbb{R}$ with right-unbounded support and finite mean on the positive half line. We say that F is *strong subexponential*, and write $F \in S^*$, if

$$\int_0^x \overline{F}(x-y)\overline{F}(y)dy \sim 2m\overline{F}(x) \quad \text{as } x \to \infty,$$

where $m = \mathbb{E}\xi^+$ and ξ has distribution F.

It follows from the observation (3.12) that a distribution F on $\mathbb{R}$ belongs to the class S^* if and only if it is heavy-tailed and sufficiently regular that

$$\lim_{x\to\infty} \frac{1}{\overline{F}(x)} \int_0^x \overline{F}(x-y)\overline{F}(y)dy$$

exists. Thus it is again the case that most heavy-tailed distributions likely to be of use in practical applications belong to the class $\mathcal{S}^*$. This includes all those named distributions introduced in Sect. 2.1, i.e. the Pareto, Burr, Cauchy, lognormal, and Weibull (with shape parameter $\alpha < 1$) distributions.

We shall see in Sect. 4.10 that the condition $F \in \mathcal{S}^*$ is equivalent to the requirement that the density f on $\mathbb{R}^+$ given by $f(x) := \overline{F}(x)/m$ be subexponential in the sense defined there.

We show first that the class $\mathcal{S}^*$ of strong subexponential distributions is a subclass of the class $\mathcal{L}$ of long-tailed distributions on $\mathbb{R}$.

Theorem 3.23. *Let the distribution F on $\mathbb{R}$ belong to $\mathcal{S}^*$. Then F is long-tailed.*

Proof. Since, for $x \geq 2$,

$$\int_0^x \overline{F}(x-y)\overline{F}(y)dy \geq 2\overline{F}(x)\int_0^1 \overline{F}(y)dy + 2\overline{F}(x-1)\int_1^{x/2}\overline{F}(y)dy,$$

the inclusion $F \in \mathcal{S}^*$ implies

$$(\overline{F}(x-1)-\overline{F}(x))\int_1^{x/2}\overline{F}(y)dy \leq \frac{1}{2}\int_0^x \overline{F}(x-y)\overline{F}(y)dy - \overline{F}(x)\int_0^{x/2}\overline{F}(y)dy$$
$$= m\overline{F}(x) - m\overline{F}(x) + o(\overline{F}(x)) \qquad \text{as } x \to \infty,$$

where again $m = \int_0^\infty \overline{F}(y)dy$. It thus follows from Lemma 2.22 that F is long-tailed. □

We now have the following analogue to the conditions for subexponentiality given by Theorem 3.7.

Theorem 3.24. *Let F be a distribution on $\mathbb{R}$. Then the following are equivalent:*
(i) $F \in \mathcal{S}^*$.
(ii) *F is long-tailed, and for every function h with $h(x) < x/2$ for all x and such that $h(x) \to \infty$ as $x \to \infty$,*

$$\int_{h(x)}^{x-h(x)} \overline{F}(x-y)\overline{F}(y)dy = o(\overline{F}(x)) \quad \text{as } x \to \infty. \qquad (3.13)$$

(iii) *There exists a function h with $h(x) < x/2$ for all x, such that $h(x) \to \infty$ as $x \to \infty$ and F is h-insensitive, and the relation (3.13) holds.*

Note that it follows in particular from Theorem 3.24 that membership of the class $\mathcal{S}^*$ is a tail property.

Proof. Note first that each of the conditions (i)–(iii) implies that F is long-tailed. This follows in the case of (i) from Theorem 3.23, and in the case of (iii) from the existence of an increasing function with respect to which F is h-insensitive. Hence we assume without loss of generality that F is long-tailed ($F \in \mathcal{L}$).

Let h be any function with $h(x) < x/2$, such that $h(x) \to \infty$ as $x \to \infty$ and F is h-insensitive. (Note as usual that since F is assumed long-tailed there exists at least one such function h.) Then, for any $x \geq 0$,

$$\int_0^x \overline{F}(x-y)\overline{F}(y)\,dy = 2\int_0^{h(x)} \overline{F}(x-y)\overline{F}(y)\,dy + \int_{h(x)}^{x-h(x)} \overline{F}(x-y)\overline{F}(y)\,dy.$$

The h-insensitivity of F implies that

$$\int_0^{h(x)} \overline{F}(x-y)\overline{F}(y)\,dy \sim m\overline{F}(x) \quad \text{as } x \to \infty,$$

where again $m = \mathbb{E}\xi^+$ and ξ has distribution F. It thus follows that the condition $F \in \mathcal{S}^*$ is equivalent to (3.13). The theorem now follows on noting that, as in the proof of Theorem 3.7, the reason why, in the statement (ii), we do not require any restriction to functions h such that F is h-insensitive follows from Proposition 2.20(ii). $\square$

We now have the following theorem and its important corollary.

Theorem 3.25. *Suppose that $F \in \mathcal{S}^*$, that G is long-tailed, and that F and G are weakly tail-equivalent. Then $G \in \mathcal{S}^*$.*

Proof. Let h be a function such that $h(x) < x/2$, $h(x) \to \infty$ and G is h-insensitive. Then, from Theorem 3.24 and the given weak tail-equivalence,

$$\int_{h(x)}^{x-h(x)} \overline{G}(x-y)\overline{G}(y)\,dy = O\left(\int_{h(x)}^{x-h(x)} \overline{F}(x-y)\overline{F}(y)\,dy\right)$$
$$= o(\overline{F}(x))$$
$$= o(\overline{G}(x)) \quad \text{as } x \to \infty.$$

Again from Theorem 3.24, it now follows that $G \in \mathcal{S}^*$. $\square$

Corollary 3.26. *Let distributions F and G be proportionally tail-equivalent. If $F \in \mathcal{S}^*$ then $G \in \mathcal{S}^*$.*

The following theorem asserts in particular that $\mathcal{S}^*$ is a subclass of $\mathcal{S}_{\mathbb{R}}$.

Theorem 3.27. *If $F \in \mathcal{S}^*$, then $F \in \mathcal{S}_{\mathbb{R}}$ and $F_I \in \mathcal{S}$.*

We do not provide a proof for this result now. Instead of that we recall the notion of an integrated weighted tail distribution and state sufficient conditions for its tail to be subexponential. Then Theorem 3.27 is a particular case of Theorem 3.28.

Let F be a distribution on $\mathbb{R}$ and let μ be a non-negative measure on $\mathbb{R}^+$ such that

$$\int_0^\infty \overline{F}(t)\,\mu(dt) \quad \text{is finite.} \tag{3.14}$$

Then, as in (2.25), we can define the distribution F_μ on $\mathbb{R}^+$ by its tail:

$$\overline{F}_\mu(x) := \min\left(1, \int_0^\infty \overline{F}(x+t)\,\mu(dt)\right), \quad x \geq 0. \tag{3.15}$$

We may now ask the following question: what type of conditions on F imply the subexponentiality of F_μ?

For any $b > 0$, define the class $\mathcal{M}_b$ of all non-negative measures μ on $\mathbb{R}^+$ such that $\mu(x, x+1] \leq b$ for all x.

Theorem 3.28. *Let $F \in \mathcal{S}^*$ and $\mu \in \mathcal{M}_b$, $b \in (0, \infty)$. Then $F_\mu \in \mathcal{S}$. Moreover, $\overline{F_\mu * F_\mu}(x) \sim 2\overline{F}_\mu(x)$ as $x \to \infty$ uniformly in $\mu \in \mathcal{M}_b$.*

Here are two examples of such measures μ: (i) if $\mu(B) = \mathbb{I}\{0 \in B\}$, then F_μ is F restricted to $\mathbb{R}^+$; (ii) if $\mu(dt) = dt$ is Lebesgue measure on $\mathbb{R}^+$, then $F_\mu = F_I$. These examples give a proof of Theorem 3.27.

Proof. First, recall that Theorem 2.28 states that if F is long-tailed, then F_μ is long-tailed uniformly in $\mu \in \mathcal{M}_b$. Thus, it is sufficient to show that, for any $h(x) \to \infty$,

$$\lim_{x \to \infty} \sup_{\mu \in \mathcal{M}_b} \frac{1}{\overline{F}_\mu(x)} \int_{h(x)}^{x-h(x)} \overline{F}_\mu(x-y) F_\mu(dy) = 0, \qquad (3.16)$$

see Theorem 3.7. For any $\mu \in \mathcal{M}_b$,

$$F_\mu(y, y+1] \leq \int_y^{y+1} \overline{F}(t) \mu(dt) \leq \overline{F}(y) \mu(y, y+1] \leq b\overline{F}(y).$$

Therefore, (3.16) holds if and only if

$$\lim_{x \to \infty} \sup_{\mu \in \mathcal{M}_b} \frac{1}{\overline{F}_\mu(x)} \int_{h(x)}^{x-h(x)} \overline{F}_\mu(x-y) \overline{F}(y) dy = 0. \qquad (3.17)$$

Since $F \in \mathcal{S}^*$, as $x \to \infty$,

$$\int_{h(x)}^{x-h(x)} \overline{F}(x-u) \overline{F}(u) du = o(\overline{F}(x)),$$

by Theorem 3.24. Then,

$$\int_{h(x)}^{x-h(x)} \overline{F}_\mu(x-y) \overline{F}(y) dy = \int_{h(x)}^{x-h(x)} \left(\int_0^\infty \overline{F}(x+t-y) \mu(dt) \right) \overline{F}(y) dy$$

$$\leq \int_0^\infty \left(\int_{h(x)}^{x+t-h(x)} \overline{F}(x+t-y) \overline{F}(y) dy \right) \mu(dt)$$

$$= \int_0^\infty o(\overline{F}(x+t)) \mu(dt) = o(\overline{F}_\mu(x))$$

and we get (3.17). $\qquad\qquad\square$

3.5 Sufficient Conditions for Subexponentiality

We formulate and prove here two results. The first may be applied to very heavy distributions such as Pareto distributions, while the second one may be applied to lighter distributions of the Weibull-type.

Theorem 3.29. *Let F be a long-tailed distribution on $\mathbb{R}$ ($F \in \mathcal{L}$) and suppose that there exists $c > 0$ such that $\overline{F}(2x) \geq c\overline{F}(x)$ for all x (that is, F belongs to the class $\mathcal{D}$ of dominated-varying distributions introduced in Sect. 2.1). Then:*
 (i) *F is whole-line subexponential.*
 (ii) *$F \in \mathcal{S}^*$, provided F has a finite mean on the positive half line.*
(iii) *$F_\mu \in \mathcal{S}$, for all μ satisfying (3.14).*

Note in particular that the statement (i) of Theorem 3.29 asserts that $\mathcal{D} \cap \mathcal{L} \subseteq \mathcal{S}_\mathbb{R}$.

Proof. It follows from the comment after Theorem 3.28 that (iii) implies (i) and (ii). Thus it is sufficient to prove (iii). The inequality $\overline{F}(2x) \geq c\overline{F}(x)$ yields, for those values of x such that the integrals below are less than 1,

$$\overline{F}_\mu(2x) = \int_0^\infty \overline{F}(2x+y)\mu(dy)$$

$$\geq c\int_0^\infty \overline{F}(x+y/2)\mu(dy)$$

$$\geq c\int_0^\infty \overline{F}(x+y)\mu(dy) = c\overline{F}_\mu(x). \qquad (3.18)$$

Now let h be any function such that $h(x) < x/2$ and $h(x) \to \infty$. We have the following bound:

$$\int_{h(x)}^{x-h(x)} \overline{F}_\mu(x-y)F_\mu(dy) = \int_{h(x)}^{x/2} \overline{F}_\mu(x-y)F_\mu(dy) + \int_{x/2}^{x-h(x)} \overline{F}_\mu(x-y)F_\mu(dy)$$
$$\leq \overline{F}_\mu(x/2)\overline{F}_\mu(h(x)) + \overline{F}_\mu(h(x))\overline{F}_\mu(x/2).$$

Therefore, by (3.18),

$$\int_{h(x)}^{x-h(x)} \overline{F}_\mu(x-y)F_\mu(dy) \leq 2\overline{F}_\mu(x)\overline{F}_\mu(h(x))/c = o(\overline{F}_\mu(x)) \quad \text{as } x \to \infty.$$

Applying now Theorem 3.7(ii), we conclude that $F_\mu \in \mathcal{S}$. $\qquad\square$

The Pareto distribution, and more generally any regularly varying or indeed intermediate regularly varying distribution, satisfies the conditions of Theorem 3.29 (i.e. belongs to $\mathcal{D} \cap \mathcal{L}$) and is, therefore, subexponential. All of the above distributions whose means are finite also belong to the to class $\mathcal{S}^*$.

However, the lognormal distribution and the Weibull distribution do not satisfy the conditions of Theorem 3.29 and we need a different technique for proving their subexponentiality.

Recall that we denote by R the hazard function given by $R(x) := -\ln \overline{F}(x)$ and by r the hazard rate function given by $r(x) = R'(x)$, provided the hazard function is differentiable.

Theorem 3.30. *Let F be a long-tailed distribution on $\mathbb{R}$ ($F \in \mathcal{L}$). Assume that there exist $\gamma < 1$ and $A < \infty$ such that the hazard function $R(x)$ satisfies the following inequality:*

$$R(x) - R(x-y) \le \gamma R(y) + A, \tag{3.19}$$

for all $x > 0$ and $y \in [0, x/2]$. If the function $e^{-(1-\gamma)R(x)}$ is integrable over $\mathbb{R}^+$, then $F \in \mathcal{S}^*$. In particular, F is whole-line subexponential ($F \in \mathcal{S}_{\mathbb{R}}$).

Proof. For any $h < x/2$,

$$\int_h^{x-h} \overline{F}(x-y)\overline{F}(y)dy = 2\int_h^{x/2} \overline{F}(x-y)\overline{F}(y)dy$$
$$= 2\overline{F}(x) \int_h^{x/2} e^{R(x)-R(x-y)-R(y)}dy.$$

It follows from (3.19) that

$$\int_h^{x/2} e^{R(x)-R(x-y)-R(y)}dy \le e^A \int_h^\infty e^{-(1-\gamma)R(y)}dy \to 0 \quad \text{as } h \to \infty,$$

since the function $e^{-(1-\gamma)R(x)}$ is integrable. Hence, if h is now any function such that $h(x) \to \infty$ then

$$\int_{h(x)}^{x-h(x)} \overline{F}(x-y)\overline{F}(y)dy = o(\overline{F}(x)).$$

Hence, by Theorem 3.24, we have $F \in \mathcal{S}^*$.   □

We note briefly that, for $0 < \gamma < 1$, the function $e^{-(1-\gamma)R(x)}$ is integrable if F has a finite moment of order $\frac{1}{1-\gamma} + \varepsilon$ on the positive half line $\mathbb{R}^+$ for some $\varepsilon > 0$. To see this note that the tail of F may then be bounded from above by $cx^{-1/(1-\gamma)-\varepsilon}$ (by the Chebyshev inequality):

$$e^{-(1-\gamma)R(x)} = (\overline{F}(x))^{1-\gamma} \le c^{1-\gamma}x^{-1-(1-\gamma)\varepsilon}.$$

The heavy-tailed Weibull distribution, with tail function $\overline{F}$ given by $\overline{F}(x) = e^{-x^\alpha}$ for some $\alpha \in (0,1)$, satisfies the conditions of Theorem 3.30. Indeed, since the function $R(x) = x^\alpha$ is concave for $\alpha \in (0,1)$, we have for $y \le x/2$ (so that $x - y \ge y$ and $R'(x-y) \le R'(y)$)

$$R(x) - R(x-y) \le yR'(x-y) \le yR'(y) = \alpha R(y).$$

Similarly, it may be checked that the lognormal distribution satisfies conditions of Theorem 3.30 and, therefore, belongs to the class $\mathcal{S}^*$.

3.6 Conditions for Subexponentiality in Terms of Truncated Exponential Moments

Note that some heavy-tailed distributions, for example those with tail functions of the form $e^{-x/\log x}$ do not satisfy the conditions of Theorem 3.30 (in this case $\gamma = 1$) and we need a more advanced technique for proving the subexponentiality of such distributions. The next two theorems, due to Pitman [45], relate the classes $\mathcal{S}$ and $\mathcal{S}^*$ to the asymptotic behaviour of truncated exponential moments with special indices.

Theorem 3.31. *Let F be a distribution on $\mathbb{R}^+$. Suppose that the hazard rate function r exists, is eventually non-increasing and that $r(x) \to 0$ as $x \to \infty$. Then F is subexponential ($F \in \mathcal{S}$) if and only if*

$$\int_0^x e^{yr(x)} F(dy) \to 1 \quad as\ x \to \infty. \tag{3.20}$$

Further, a sufficient condition for subexponentiality is that the function of y given by $e^{yr(y)-R(y)} r(y)$ is integrable over $\mathbb{R}^+$. Here $R(x) = \int_0^x r(y) dy$ defines the corresponding hazard function.

Note that the integral in (3.20) is equal to $\mathbb{E}\{e^{\xi r(x)}; \xi \leq x\}$, where ξ is a random variable with distribution F.

Proof. The proof consists of two steps. First, we show that it may be assumed without loss of generality that the function $r(x)$ is non-increasing for all x (and that the condition (3.20) is a *tail property*), and then we prove the results under this assumption.

Suppose first that $r(x)$ may increase in a neighbourhood of 0 but is non-increasing for all $x \geq x_*$. Define the non-increasing hazard rate

$$r_*(x) = \begin{cases} r(x_*) & \text{for } x \leq x_*, \\ r(x) & \text{for } x > x_*, \end{cases}$$

and put $R_*(x) = \int_0^x r_*(y) dy$. Define also the distribution F_* by $\overline{F}_*(x) = e^{-R_*(x)}$. Then, for all $x \geq x_*$,

$$R_*(x) - R(x) = \int_0^{x_*} (r(x_*) - r(y)) dy =: c_*,$$

and hence, by Theorem 3.11, either both F_* and F are subexponential or both are not. We now prove that the functions r and r_* either both satisfy, or else both fail to satisfy, the condition (3.20). Note first that $\overline{F}_*(x_*) = e^{-c_*} \overline{F}(x_*)$. Note that $r(x) \to 0$ as $x \to \infty$ implies that $e^{yr(x)} \to 1$ as $x \to \infty$, uniformly in $0 \leq y \leq x_*$. Hence

$$\int_0^x e^{yr(x)} F(dy) = \left(\int_0^{x_*} + \int_{x_*}^x \right) e^{yr(x)} F(dy)$$

$$= F[0, x_*] + o(1) + \int_{x_*}^x e^{yr(x)} F(dy).$$

It now follows that F satisfies the condition (3.20) if and only if

$$- \int_{x_*}^x e^{yr(x)} F(dy) = \int_{x_*}^x F(dy) + o(1) = \overline{F}(x_*) + o(1).$$

Note also that $F_*(dx) = e^{-c_*} F(dx)$ for $x > x_*$. Therefore, as $x \to \infty$,

$$\int_0^x e^{yr_*(x)} F_*(dy) = \int_0^{x_*} e^{yr_*(x)} F_*(dy) + \int_{x_*}^x e^{yr(x)} F_*(dy)$$

$$= F_*[0,x_*] + o(1) + e^{-c_*} \int_{x_*}^x e^{yr(x)} F(dy)$$

$$= F_*[0,x_*] + o(1) + e^{-c_*} (\overline{F}(x_*) + o(1))$$

$$= F_*[0,x_*] + \overline{F}_*(x_*) + o(1) = 1 + o(1).$$

where the equality in the second line follows since $r(x) \to 0$ and from the definitions, and equality in the third line holds if and only if and only if F satisfies (3.20). Thus, we have shown that F satisfies (3.20) if and only if F_* satisfies the analogous condition, with F_* in place of F and r_* in place of r. In other words, without loss of generality, we may assume from the very beginning that $r(x)$ is non-increasing for all $x \geq 0$.

It follows from the definition (3.1) that subexponentiality is equivalent to the convergence: as $x \to \infty$,

$$\int_0^x e^{R(x)-R(x-y)} F(dy) = \int_0^x e^{R(x)-R(x-y)-R(y)} r(y) dy \to 1. \qquad (3.21)$$

Since $r(x) = R'(x)$ is non-increasing, $R(x)$ is concave and

$$R(x) - R(x-y) \geq yr(x) \quad \text{for any } y \in [0,x].$$

Hence, subexponentiality in the form (3.21) implies

$$\limsup_{x \to \infty} \int_0^x e^{yr(x)} F(dy) \leq 1.$$

Together with the fact that the integral in the above expression is at least $F[0,x]$, this implies (3.20).

Now suppose that (3.20) holds. We make use of the following representation:

$$\int_0^x e^{R(x)-R(x-y)} F(dy) = \left(\int_0^{x/2} + \int_{x/2}^x \right) e^{R(x)-R(x-y)-R(y)} r(y) dy$$

$$= \int_0^{x/2} e^{R(x)-R(x-y)-R(y)} r(y) dy$$

$$+ \int_0^{x/2} e^{R(x)-R(x-y)-R(y)} r(x-y) dy$$

$$=: I_1 + I_2.$$

The first integral is not less than $F[0, x/2]$ which tends to 1 as $x \to \infty$. On the other hand, for $y \leq x/2$, and therefore $x - y \geq x/2$,

$$R(x) - R(x - y) \leq yr(x - y) \leq yr(x/2). \tag{3.22}$$

Thus,

$$I_1 \leq \int_0^{x/2} e^{yr(x/2)} F(dy),$$

which tends to 1 as $x \to \infty$ by (3.20). Thus $I_1 \to 1$.

On noting that, for any fixed y, $e^{R(x)-R(x-y)-R(y)}r(y) \to e^{-R(y)}r(y)$ as $x \to \infty$ and that $\int_0^\infty e^{-R(y)}r(y)dy = 1$, we obtain that the family (in x) of functions (in y)

$$z_x(y) = e^{R(x)-R(x-y)-R(y)}r(y)\mathbb{I}\{y \leq x/2\}$$

is uniformly integrable in the sense that

$$\sup_x \int_A^\infty z_x(y)dy \to 0 \quad \text{as } A \to \infty.$$

Since $r(x - y) \leq r(y)$ for all $y \leq x/2$, the integrand in I_2 is dominated by $z_x(y)$. It follows that $I_2 \to 0$ as $x \to \infty$, since also $e^{R(x)-R(x-y)-R(y)}r(x - y) \leq r(x - y) \to 0$ for any fixed y. Thus (3.21) holds; that is, the condition (3.20) implies subexponentiality.

The second part of the theorem follows by dominated convergence, since, for all sufficiently large $y < x$, we have $r(x) \leq r(y)$. $\qquad \square$

As an example, consider a distribution F such that, for some $\alpha > 0$ and for all sufficiently large x,

$$\overline{F}(x) = e^{-x/\log^\alpha x} \tag{3.23}$$

Then, again for sufficiently large x, the hazard rate function r is given by $r(x) = 1/\log^\alpha x - \alpha/\log^{\alpha+1} x$ and the function

$$e^{xr(x)-R(x)}r(x) = e^{-\alpha x/\log^{\alpha+1} x}r(x)$$

(where, as usual, R is the corresponding hazard function) is integrable over $\mathbb{R}^+$. Therefore, by Theorem 3.31, F is subexponential.

In the following theorem we give an applicable necessary and sufficient condition for membership of the class $\mathcal{S}^*$.

Theorem 3.32. *Let F be a distribution on $\mathbb{R}^+$ with finite mean m. Suppose that the hazard rate function r exists, is eventually non-increasing, and that $r(x) \to 0$ as $x \to \infty$. Then F is strong subexponential if and only if*

$$\int_0^x e^{yr(x)}\overline{F}(y)dy \to m \quad \text{as } x \to \infty. \tag{3.24}$$

Further, a sufficient condition for $F \in \mathcal{S}^*$ is that (the function of y given by) $e^{yr(y)}\overline{F}(y)$ is integrable over $\mathbb{R}^+$.

Proof. Arguments similar to those used in the proof of Theorem 3.31 show that without loss of generality we may assume that the corresponding hazard function R satisfies $R(0) = 0$ and that the hazard rate function r is non-increasing over all of $\mathbb{R}^+$. The distribution F belongs to the class $\mathcal{S}^*$ if and only if, as $x \to \infty$,

$$\int_0^{x/2} e^{R(x)-R(x-y)}\overline{F}(y)dy := \int_0^{x/2} e^{R(x)-R(x-y)-R(y)}dy \to m. \qquad (3.25)$$

Since $r(x) = R'(x)$ is non-increasing, $R(x)$ is concave and

$$R(x) - R(x-y) \geq yr(x) \quad \text{for any } y \in [0,x].$$

Suppose first that the condition (3.25) holds. Then

$$\limsup_{x \to \infty} \int_0^x e^{yr(x)}\overline{F}(y)dy \leq m.$$

However, we also have that the latter integral is at least $\int_0^x \overline{F}(y)dy$ which tends to m as $x \to \infty$. Hence the condition (3.24) follows.

Now suppose instead that the condition (3.24) holds. It follows from (3.22) that

$$\int_0^{x/2} e^{R(x)-R(x-y)}\overline{F}(y)dy \leq \int_0^{x/2} e^{yr(x/2)}\overline{F}(y)dy,$$

and from (3.24) that this latter integral tends to m as $x \to \infty$.

The second part of the theorem follows by dominated convergence, since, again for all sufficiently large $y < x$, we have $r(x) \leq r(y)$. $\qquad \square$

As an example we again consider a distribution F whose tail is such that, for some $\alpha > 0$ and for all sufficiently large x, the relation (3.23) holds. We now have that $F \in \mathcal{S}^*$, since in this case the function

$$e^{xr(x)-R(x)} = e^{-\alpha x/\log^{\alpha+1} x}$$

(where the hazard rate function r is given as previously and R is again the corresponding hazard function) is integrable over $\mathbb{R}^+$.

3.7 $\mathcal{S}$ Is a Proper Subset of $\mathcal{L}$

In this section we use Theorem 3.31 to construct a distribution F which is long-tailed but not subexponential. Fix any decreasing sequence $\alpha_n \to 0$ as $n \to \infty$. The corresponding hazard function $R(x)$ will be defined as continuous and piecewise linear so that the hazard rate function $r(x) := R'(x) = \alpha_n$ for $x \in (x_{n-1},x_n]$. Since on the interval $y \in (x_{n-1},x_n]$

$$yr(x_n) - R(y) = y\alpha_n - [R(x_{n-1}) + \alpha_n(y - x_{n-1})] > -R(x_{n-1}),$$

we have the following lower bound for the integral on the left side of (3.20):

$$\int_{x_{n-1}}^{x_n} e^{yr(x_n)-R(y)} r(y) dy \geq \int_{x_{n-1}}^{x_n} e^{-R(x_{n-1})} \alpha_n dy$$
$$= \alpha_n (x_n - x_{n-1}) e^{-R(x_{n-1})}.$$

Now choose $x_0 = 0$, $R(x_0) = 0$, and the x_n so that

$$\alpha_n (x_n - x_{n-1}) e^{-R(x_{n-1})} = 2.$$

For this we take $x_n = x_{n-1} + 2\alpha_n^{-1} e^{R(x_{n-1})}$ and then

$$R(x_n) = R(x_{n-1}) + \alpha_n (x_n - x_{n-1})$$
$$= R(x_{n-1}) + 2e^{R(x_{n-1})}.$$

Clearly $R(x) \to \infty$ as $x \to \infty$. Since $\alpha_n \to 0$ as $n \to \infty$, $r(x) \to 0$ as $x \to \infty$; thus, $\overline{F}(x) = e^{-R(x)}$ is long-tailed (see Sect. 2.5). On the other hand, by the above construction,

$$\int_{x_{n-1}}^{x_n} e^{yr(x_n)-R(y)} r(y) dy \geq 2 \quad \text{for all } n,$$

so that

$$\int_0^{x_n} e^{yr(x_n)-R(y)} r(y) dy$$

does not converge to 1 as $n \to \infty$. It now follows from Theorem 3.31 that F is not subexponential.

The idea in this example is that the tail $\overline{F}$ is a piecewise exponential function; the indexes of the exponents tend to zero and the lengths of the intervals of exponentiality grow very fast.

3.8 Does $F \in S$ Imply That $F_I \in S$?

It is natural to consider the following question: May the assumption $F \in S^*$ of Theorem 3.28 be weakened to $F \in S$? In the case of Lebesgue measure μ, i.e. where $F_\mu = F_I$, this question is raised in [24, Sect. 1.4.2].

In this section, we answer the above question in the negative by giving an example of a distribution $F \in S$ with finite mean such that $F_I \notin S$. This example is based on the following construction.

Define $R_0 = 0$, $R_1 = 1$ and $R_{n+1} = e^{R_n}/R_n$. Since e^x/x is increasing for $x \geq 1$, the sequence R_n is increasing and

$$R_n = o(R_{n+1}) \quad \text{as } n \to \infty. \tag{3.26}$$

Put $t_n = R_n^2$. Define the hazard function $R(x) := -\ln \overline{F}(x)$ as

$$R(x) = R_n + r_n(x - t_n) \quad \text{for } x \in (t_n, t_{n+1}],$$

where

$$r_n = \frac{R_{n+1} - R_n}{t_{n+1} - t_n} = \frac{1}{R_{n+1} + R_n} \tag{3.27}$$

$$\sim \frac{1}{R_{n+1}} \qquad \text{as } n \to \infty \tag{3.28}$$

by (3.26). In other words, the hazard rate $r(x) = R'(x)$ is defined as $r(x) = r_n$ for $x \in (t_n, t_{n+1}]$, where r_n is given by (3.27). By construction, we have

$$\overline{F}(t_n) = e^{-\sqrt{t_n}},$$

so that at the points t_n the tail function $\overline{F}$ of the distribution F behaves like that of the Weibull distribution with parameter $1/2$. Between these points the tail decays exponentially with indexes r_n.

We shall prove that $F \in \mathcal{S}$ and has finite mean, but that $F_I \notin \mathcal{S}$. Let

$$J_n := F_I(t_n, t_{n+1}] = \int_{t_n}^{t_{n+1}} \overline{F}(u)du = \int_{t_n}^{t_{n+1}} e^{-R(u)}du.$$

Since by (3.28)

$$J_n = r_n^{-1}\left(e^{-R_n} - e^{-R_{n+1}}\right)$$
$$\sim r_n^{-1}e^{-R_n} \sim R_{n+1}e^{-R_n} = 1/R_n \qquad \text{as } n \to \infty, \tag{3.29}$$

the mean of F,

$$\int_0^\infty \overline{F}(y)dy = \sum_{n=0}^\infty \int_{t_n}^{t_{n+1}} \overline{F}(y)dy,$$

is finite.

It follows from (3.27) that $r(x)$ is eventually decreasing and tends to 0, and we can thus apply Theorem 3.31 to show that F is subexponential. By that theorem, F is subexponential provided the function $e^{yr(y)-R(y)}r(y)$ is integrable over $\mathbb{R}^+$. We estimate the integral of this function. Put

$$I_n = \int_{t_n}^{t_{n+1}} e^{yr(y)-R(y)}r(y)dy.$$

Then

$$I_n = r_n \int_{t_n}^{t_{n+1}} e^{yr_n - R_n - r_n(y - t_n)}dy \leq r_n e^{-R_n + r_n t_n} t_{n+1}.$$

Since, as $n \to \infty$,

$$r_n t_{n+1} = r_n R_{n+1}^2 \sim R_{n+1} \tag{3.30}$$

by (3.28) and

$$r_n t_n = r_n R_n^2 \sim R_n^2/R_{n+1} = R_n^3 e^{-R_n} \to 0, \tag{3.31}$$

we get, for n sufficiently large,

$$I_n \le 2R_{n+1}e^{-R_n} \sim 2/R_n.$$

Therefore,

$$\int_0^\infty e^{yr(y)-R(y)}r(y)dy = \sum_{n=0}^\infty I_n < \infty,$$

and F is indeed subexponential.

In order to prove that F_I is not subexponential, we again take use of Theorem 3.31. It suffices to prove that

$$\int_0^x e^{yr(x)}F_I(dy) = \int_0^x e^{yr(x)}\overline{F}(y)dy \to \infty$$

for some subsequence of points x. If we take $x = t_{n+1}$, then the latter integral is not less than

$$\int_{t_n}^{t_{n+1}} e^{yr(t_{n+1})}\overline{F}(y)dy = \int_{t_n}^{t_{n+1}} e^{yr_n}e^{-R_n-r_n(y-t_n)}dy$$
$$\ge e^{-R_n}(t_{n+1}-t_n).$$

Then, for all those n where $t_{n+1} > 2t_n$,

$$\int_0^{t_{n+1}} e^{yr(t_{n+1})}F_I(dy) \ge e^{-R_n}t_{n+1}/2 = e^{-R_n}R_{n+1}^2/2 = e^{R_n}/2R_n^2.$$

which tends to infinity as $n \to \infty$. so that F_I is not subexponential. Thus, $F \in \mathcal{S}$ and has finite mean, but $F_I \notin \mathcal{S}$.

3.9 Closure Properties of the Class of Subexponential Distributions

In this section, we discuss the following question: is the class $\mathcal{S}_\mathbb{R}$ closed under convolution? It is well-known that the class of regularly varying distributions, which is a subclass of the class $\mathcal{S}_\mathbb{R}$ of subexponential distributions, is closed under convolution. Indeed if F and G are regularly varying, the result that $F * G$ is also regularly varying is straightforwardly obtained from Theorem 3.14 by taking the "reference" distribution of that theorem to be $(F + G)/2$. It is also known that the class $\mathcal{S}_\mathbb{R}$ does not possess this closure property. However, if distributions F, $G \in \mathcal{S}_\mathbb{R}$, then it follows from Corollary 3.16 that a sufficient condition for $F * G \in \mathcal{S}_\mathbb{R}$ is given by $\overline{G}(x) = O(\overline{F}(x))$ as $x \to \infty$. (Indeed, as the corollary shows, G may satisfy weaker conditions than that of being subexponential.) Further it follows that under this condition we have that, for any function h such that $h(x) \to \infty$ and both F and G are h-insensitive,

$$\mathbb{P}\{\xi+\eta > x, \xi > h(x), \eta > h(x)\} = o(\overline{F}(x)+\overline{G}(x)) \quad \text{as } x \to \infty, \qquad (3.32)$$

where ξ and η are independent random variables with respective distributions F and G. (See, for example, the proof of Theorem 3.14 above.) The following result is therefore not surprising: if $F, G \in \mathcal{S}_{\mathbb{R}}$, the condition (3.32) is *necessary and sufficient* for $F * G \in \mathcal{S}_{\mathbb{R}}$.

Theorem 3.33. *Suppose that the distributions F and G on $\mathbb{R}$ are subexponential. Then the following conditions are equivalent:*
  (i) $\overline{F * G}(x) \sim \overline{F}(x) + \overline{G}(x)$ *as* $x \to \infty$
  (ii) $F * G \in \mathcal{S}_{\mathbb{R}}$
  (iii) *The mixture* $pF + (1 - p)G$ *belongs to* $\mathcal{S}_{\mathbb{R}}$ *for all p satisfying* $0 < p < 1$
  (iv) *The mixture* $pF + (1 - p)G$ *belongs to* $\mathcal{S}_{\mathbb{R}}$ *for some p satisfying* $0 < p < 1$
  (v) *The relation (3.32) holds for any function h such that $h(x) \to \infty$ as $x \to \infty$ and both F and G are h-insensitive*
  (vi) *The relation (3.32) holds for some function h such that $h(x) \to \infty$ as $x \to \infty$ and both F and G are h-insensitive*

Proof. Let h be any function such that $h(x) \to \infty$ as $x \to \infty$ and both F and G are h-insensitive. We show that each of the conditions (i)–(iv) is equivalent to (3.32). The equivalence of the conditions (i)–(vi) of the theorem is then immediate. First, since F and G are subexponential, and hence long-tailed, it follows from the decomposition (2.18) and Lemma 2.34 that

$$\overline{F * G}(x)$$
$$= \overline{F}(x) + o(\overline{F}(x)) + \overline{G}(x) + o(\overline{G}(x)) + \mathbb{P}\{\xi + \eta > x, \xi > h(x), \eta > h(x)\}. \quad (3.33)$$

Hence the condition (i) and (3.32) are equivalent.

To show the equivalence of (ii) and (3.32) observe first that subexponentiality of F and G implies that

$$\overline{F^{*2}}(x) \sim 2\overline{F}(x), \quad \overline{G^{*2}}(x) \sim 2\overline{G}(x), \quad (3.34)$$

and thus in particular, from Lemma 2.37, that

$$\mathbb{P}\{\xi_1 + \xi_2 + \eta_1 + \eta_2 > x, \xi_1 + \xi_2 > h(x), \eta_1 + \eta_2 > h(x)\}$$
$$\sim 4\mathbb{P}\{\xi + \eta > x, \xi > h(x), \eta > h(x)\}. \quad (3.35)$$

Further, since $(F * G)^{*2} = F^{*2} * G^{*2}$ and since both F^{*2} and G^{*2} are h-insensitive, $\overline{(F * G)^{*2}}(x)$ may be estimated as in (3.33) with F^{*2} and G^{*2} replacing F and G. Hence, using also (3.34) and (3.35),

$$\overline{(F * G)^{*2}}(x)$$
$$= (2 + o(1))(\overline{F}(x) + \overline{G}(x)) + (4 + o(1))\mathbb{P}\{\xi + \eta > x, \xi > h(x), \eta > h(x)\}. \quad (3.36)$$

Now since subexponentiality of F and G also implies, by Corollary 2.42, that $F * G \in \mathcal{L}$, the condition (ii) is equivalent to the requirement that

$$\overline{(F*G)^{*2}}(x) = (2+o(1))\overline{F*G}(x)$$
$$= (2+o(1))(\overline{F}(x)+\overline{G}(x)) + (2+o(1))\mathbb{P}\{\xi+\eta > x, \xi > h(x), \eta > h(x)\},$$

$$(3.37)$$

where (3.37) follows from (3.33). However, the equalities (3.36) and (3.37) hold simultaneously if and only if (3.32) holds.

Finally, to show the equivalence of (iii) (and (iv)) and (3.32), fix p such that $0 < p < 1$ and note first that $pF + (1-p)G$ is h-insensitive. Hence, by Theorem 3.6, subexponentiality of $pF + (1-p)G$ is equivalent to

$$\int_{h(x)}^{\infty} \overline{pF+(1-p)G}(\max(h(x), x-y))(pF+(1-p)G)(dy) = o(\overline{F}(x)+\overline{G}(x)).$$

The left side is equal to

$$p^2\mathbb{P}\{\xi_1+\xi_2 > x, \xi_1 > h(x), \xi_2 > h(x)\} + (1-p)^2\mathbb{P}\{\eta_1+\eta_2 > x, \eta_1 > h(x), \eta_2 > h(x)\}$$
$$+ 2p(1-p)\mathbb{P}\{\xi+\eta > x, \xi > h(x), \eta > h(x)\}.$$

By subexponentiality of F and G and again by Theorem 3.6, $\mathbb{P}\{\xi_1 + \xi_2 > x, \xi_1 > h(x), \xi_2 > h(x)\} = o(\overline{F}(x))$ and $\mathbb{P}\{\eta_1 + \eta_2 > x, \eta_1 > h(x), \eta_2 > h(x)\} = o(\overline{G}(x))$. The equivalence of (iii) and (3.32) now follows. □

In general, the class $\mathcal{S}_{\mathbb{R}}$ is not closed under convolutions. An example of two subexponential distributions F_1 and F_2 such that $F_1 * F_2$ is not subexponential was constructed by Leslie in [40].

3.10 Kesten's Bound

We know that if a distribution F on $\mathbb{R}$ is subexponential ($F \in \mathcal{S}_{\mathbb{R}}$) then $\overline{F^{*n}}(x)/\overline{F}(x) \to n$ as $x \to \infty$. However, for many purposes, e.g. the application of the dominated convergence theorem, an upper bound for $\overline{F^{*n}}(x)/\overline{F}(x)$ is required. One such is given by the theorem below, known as Kesten's bound.

Theorem 3.34. *Suppose that $F \in \mathcal{S}_{\mathbb{R}}$. Then, for any $\varepsilon > 0$, there exists $c(\varepsilon) > 0$ such that, for any $x \geq 0$ and $n \geq 1$,*

$$\overline{F^{*n}}(x) \leq c(\varepsilon)(1+\varepsilon)^n \overline{F}(x).$$

Proof. Since for any random variable ξ we have $\xi \leq \xi^+$, it is sufficient to prove the theorem for distributions on the positive half-line $\mathbb{R}^+$. Let $\xi_1, \xi_2, \ldots$ be a sequence of independent random variables with common distribution F, and, for each n, let $S_n = \sum_{i=1}^{n} \xi_i$. For $x_0 > 0$ and $k \geq 1$, put

$$A_k := A_k(x_0) = \sup_{x > x_0} \frac{\overline{F^{*k}}(x)}{\overline{F}(x)}.$$

Take $\varepsilon > 0$. It follows from subexponentiality that there exists x_0 such that, for any $x > x_0$,

$$\mathbb{P}\{\xi_1 + \xi_2 > x, \ \xi_2 \leq x\} = \mathbb{P}\{\xi_1 + \xi_2 > x\} - \mathbb{P}\{\xi_2 > x\}$$
$$\leq (1 + \varepsilon/2)\overline{F}(x).$$

We have the following decomposition

$$\mathbb{P}\{S_n > x\} = \mathbb{P}\{S_n > x, \ \xi_n \leq x - x_0\} + \mathbb{P}\{S_n > x, \ \xi_n > x - x_0\}$$
$$=: P_1(x) + P_2(x).$$

By the definitions of A_{n-1} and x_0, for any $x > x_0$,

$$P_1(x) = \int_0^{x-x_0} \mathbb{P}\{S_{n-1} > x - y\}\mathbb{P}\{\xi_n \in dy\}$$
$$\leq A_{n-1}\int_0^{x-x_0} \overline{F}(x-y)\mathbb{P}\{\xi_n \in dy\}$$
$$= A_{n-1}\mathbb{P}\{\xi_1 + \xi_n > x, \ \xi_n \leq x - x_0\}$$
$$\leq A_{n-1}(1 + \varepsilon/2)\overline{F}(x). \tag{3.38}$$

Further, for any $x > x_0$,

$$P_2(x) \leq \mathbb{P}\{\xi_n > x - x_0\} \leq L\overline{F}(x), \tag{3.39}$$

where

$$L = \sup_y \frac{\overline{F}(y - x_0)}{\overline{F}(y)}.$$

Since F is long-tailed, L is finite. It follows from (3.38) and (3.39) that $A_n \leq A_{n-1}(1 + \varepsilon/2) + L$ for $n > 1$. Therefore, an induction argument yields:

$$A_n \leq A_1(1 + \varepsilon/2)^{n-1} + L\sum_{l=0}^{n-2}(1 + \varepsilon/2)^l \leq Ln(1 + \varepsilon/2)^{n-1}.$$

This implies the conclusion of the theorem. $\qquad\square$

It is straightforward to check that the above proof depends on F only through the quantity $|\overline{F * F}(x)/\overline{F}(x) - 2|$. We hence obtain immediately the following uniform version of Kesten's bound.

Theorem 3.35. *Suppose that the family of distributions $\mathcal{F}$ is uniformly subexponential, that is,*

$$\sup_{F \in \mathcal{F}} \left|\frac{\overline{F * F}(x)}{\overline{F}(x)} - 2\right| \to 0 \quad as \ x \to \infty, \tag{3.40}$$

and, in addition, for any $y > 0$,

$$\sup_{F \in \mathcal{F}} \sup_x \frac{\overline{F}(x-y)}{\overline{F}(x)} < \infty. \tag{3.41}$$

Then, for any $\varepsilon > 0$, there exists $c(\varepsilon) > 0$ such that, for any $F \in \mathcal{F}$, $x \geq 0$ and $n \geq 1$,

$$\overline{F^{*n}}(x) \leq c(\varepsilon)(1+\varepsilon)^n \overline{F}(x).$$

Recall from Sect. 3.4 that, for any $b > 0$, we define the class $\mathcal{M}_b$ to consist of all non-negative measures μ on $\mathbb{R}^+$ such that $\mu(x, x+1] \leq b$ for all x. As before, we define the distribution F_μ on $\mathbb{R}^+$ by its tail:

$$\overline{F}_\mu(x) := \min\left(1, \int_0^\infty \overline{F}(x+t)\mu(dt)\right), \quad x \geq 0.$$

We now have the following corollary to Theorem 3.35.

Corollary 3.36. *Assume that $F \in \mathcal{S}^*$ and $b > 0$. Then, for any $\varepsilon > 0$, there exists $c(\varepsilon) > 0$ such that, for all $\mu \in \mathcal{M}_b$, $x \geq 0$ and $n \geq 1$,*

$$\overline{F_\mu^{*n}}(x) \leq c(\varepsilon)(1+\varepsilon)^n \overline{F}_\mu(x).$$

Proof. We check the conditions of Theorem 3.35. The uniform subexponentiality follows from Theorem 3.28. Fix $y > 0$. Since $F \in \mathcal{S}^*$, F is long-tailed and, therefore, there exists $c < \infty$ such that $\overline{F}(x-y) \leq c\overline{F}(x)$ for all x. Then

$$\int_0^\infty \overline{F}(x-y+t)\mu(dt) \leq c \int_0^\infty \overline{F}(x+t)\mu(dt),$$

and so also the condition (3.41) holds. $\qquad\square$

3.11 Subexponentiality and Randomly Stopped Sums

In this section we study tail asymptotics for the distribution of a sum of independent identically distributed random variables stopped at a random time which is independent of the summands. These results may be used in a variety of areas including the theory of random walks, branching processes, infinitely divisible laws, etc.

Let $\xi, \xi_1, \xi_2, \ldots$ be independent random variables with a common distribution F on $\mathbb{R}^+$. Let $S_0 = 0$ and, for $n \geq 1$, let $S_n = \xi_1 + \ldots + \xi_n$. Let the counting random variable τ be independent of the sequence $\{\xi_n\}$ and take values in $\mathbb{Z}^+$. Then the distribution of S_τ is given by

$$F^{*\tau} = \sum_{n=0}^\infty \mathbb{P}\{\tau = n\}F^{*n}. \tag{3.42}$$

The first result below says that if the random variable τ has a light-tailed distribution and the distribution F of the random variables ξ_i has a subexponential distribution, then again there holds the "principle of a single big jump" introduced in Sect. 3.1.

Theorem 3.37. *Suppose that* $\mathbb{E}\tau < \infty$, *that* $F \in \mathcal{S}_{\mathbb{R}}$ *and that* $\mathbb{E}(1+\delta)^{\tau} < \infty$ *for some* $\delta > 0$. *Then*

$$\frac{\mathbb{P}\{S_{\tau} > x\}}{\overline{F}(x)} \to \mathbb{E}\tau \quad as\ x \to \infty. \tag{3.43}$$

Proof. The proof is immediate from Corollary 3.20, Theorem 3.34, and the dominated convergence theorem.   □

Here the result is valid for any subexponential distribution on the whole real line. For a fixed distribution F, the condition $\mathbb{E}(1+\delta)^{\tau} < \infty$ may be substantially weakened. We can illustrate this by the following example. Assume that there exist finite positive constants c and α such that $\overline{F}(x/n) \leq cn^{\alpha}\overline{F}(x)$ for all $x > 0$ and $n \geq 1$ (for instance, the Pareto distribution with parameter α satisfies this condition). Then $\mathbb{P}\{S_{\tau} > x\} \sim \mathbb{E}\tau \cdot \overline{F}(x)$ as $x \to \infty$ provided $\mathbb{E}\tau^{1+\alpha}$ is finite; this follows by combining the dominated convergence with the inequalities

$$\mathbb{P}\{S_n > x\} \leq \mathbb{P}\{n \cdot \max_{k \leq n}\xi_k > x\} \leq n\mathbb{P}\{\xi_1 > x/n\} \leq n^{1+\alpha}\overline{F}(x).$$

The next result shows that subexponentiality on the positive half-line $\mathbb{R}^+$ is essentially characterised by the relation (3.43).

Theorem 3.38. *Suppose that* $\mathbb{E}\tau < \infty$ *and that* $\mathbb{P}\{\tau > 1\} > 0$. *Suppose further that the distribution* F *is concentrated on* $\mathbb{R}^+$ *and that*

$$\limsup_{x \to \infty} \frac{\mathbb{P}\{S_{\tau} > x\}}{\overline{F}(x)} \leq \mathbb{E}\tau.$$

Then $F \in \mathcal{S}$.

Proof. For each positive integer k, let $p_k = \mathbb{P}\{\tau = k\}$; note also that, from (2.6), since F is concentrated on $\mathbb{R}^+$ and has unbounded support,

$$\liminf_{x \to \infty} \frac{\overline{F^{*k}}(x)}{\overline{F}(x)} \geq k. \tag{3.44}$$

Let $n \geq 2$ be such that $p_n > 0$. Then, from (3.42) and the theorem hypothesis,

$$\mathbb{E}\tau \geq \limsup_{x \to \infty} \frac{\mathbb{P}\{S_{\tau} > x\}}{\overline{F}(x)}$$

$$\geq \liminf_{x \to \infty} \sum_{k \neq n} p_k \frac{\overline{F^{*k}}(x)}{\overline{F}(x)} + p_n \limsup_{x \to \infty} \frac{\overline{F^{*n}}(x)}{\overline{F}(x)}$$

$$\geq \sum_{k \neq n} p_k \liminf_{x \to \infty} \frac{\overline{F^{*k}}(x)}{\overline{F}(x)} + p_n \limsup_{x \to \infty} \frac{\overline{F^{*n}}(x)}{\overline{F}(x)}$$

$$\geq \sum_{k \neq n} p_k k + p_n \limsup_{x \to \infty} \frac{\overline{F^{*n}}(x)}{\overline{F}(x)},$$

where the third line in the above display follows from Fatou's Lemma and the last line follows from (3.44). Since also $\mathbb{E}\tau = \sum_{k\geq 0} p_k k$ and $p_n > 0$, it follows that

$$\limsup_{x\to\infty} \frac{\overline{F^{*n}}(x)}{\overline{F}(x)} \leq n,$$

which, by Theorem 3.21, implies the subexponentiality of F. ·   □

The uniform version of Kesten's bound, see Theorem 3.35, implies the following result for families of distributions.

Theorem 3.39. *Let $\delta > 0$ and $c < \infty$. Suppose that the family of distributions $\mathcal{F}$ is uniformly subexponential, that is,*

$$\sup_{F\in\mathcal{F}} \left| \frac{\overline{F*F}(x)}{\overline{F}(x)} - 2 \right| \to 0 \quad as\ x \to \infty,$$

and that, in addition, for any $y > 0$,

$$\sup_{F\in\mathcal{F}} \sup_x \frac{\overline{F}(x-y)}{\overline{F}(x)} < \infty.$$

Then

$$\sum_{n=0}^{\infty} \overline{F^{*n}}(x)\mathbb{P}\{\tau = n\} \sim \mathbb{E}\tau\overline{F}(x)$$

as $x \to \infty$ uniformly in $F \in \mathcal{F}$ and in all τ such that $\mathbb{E}(1+\delta)^\tau \leq c$.

Together with Corollary 3.36, Theorem 3.39 implies the following uniform asymptotics.

Corollary 3.40. *Let $b > 0$, $\delta > 0$ and $c < \infty$. Suppose that $F \in \mathcal{S}^*$. Then*

$$\sum_{n=0}^{\infty} \overline{F_\mu^{*n}}(x)\mathbb{P}\{\tau = n\} \sim \mathbb{E}\tau\overline{F}_\mu(x)$$

as $x \to \infty$ uniformly in $\mu \in \mathcal{M}_b$ and in all τ such that $\mathbb{E}(1+\delta)^\tau \leq c$.

3.12 Comments

The concept of subexponential distributions (but not the name) was introduced by Chistyakov in [13], in the context of branching processes. In the same paper, the present Lemma 3.2 was established as well as some sufficient conditions for subexponentiality. Also, Theorem 3.34 (Kesten's bound) was proved under an additional technical assumption. The first book containing subexponential stuff was [7] by Athreya and Ney; to the best of our knowledge, the term subexponential class of distributions (as well as the notation $\mathcal{S}$) was first introduced in this book. This book also contains the general case of Kesten's bound (Lemma 4.4.7 there), the proof is due to Kesten, as written in the second paragraph on p. 148 in [7] and in Remark 2 on page 267 in [14] (an analogue result for the class S_γ, $\gamma > 0$, see Chover, Ney, and Wainger [15]).

The notion of weak tail-equivalence and Theorem 3.11 go back to Klüppelberg [32].

The class S^* was introduced by Klüppelberg [32].

Corollary 3.18 was proved by Embrechts, Goldie, and Veraverbeke [23].

The version of Corollary 3.16 with $G \in \mathcal{L}$ was proved by Embrechts and Goldie [21]). Corollary 3.19 is well-known (and goes back to Embrechts and Goldie [22] where the case $n = 2$, $G_1 = G_2$ was considered; some particular results may be found in Teugels [52] and Pakes [44], see also Asmussen, Foss, and Korshunov [4]).

Theorem 3.31 is due to Pitman [45].

Examples where F is long-tailed but not subexponential can be found in Embrechts and Goldie [21], and in Pitman [45]. Here we have followed the idea of Pitman.

The first four equivalences given by Theorem 3.33 were proved by Embrechts and Goldie in [21].

3.13 Problems

3.1. Prove the Cauchy distribution is subexponential and strong subexponential.

3.2. Using direct estimates for convolution of the Pareto density, show the Pareto distribution is subexponential and strong subexponential.

3.3. Using direct estimates for convolutions, prove that any regularly varying at infinity distribution is subexponential and strong subexponential.

3.4. Let F and G be two distributions that are regularly varying at infinity, and let $0 < p < 1$.

 (i) Prove that the distribution $pF + (1 - p)G$ is regularly varying at infinity.
 (ii) Prove that the distribution $F * G$ is regularly varying at infinity.

3.5. Prove by direct calculations of convolution density that the exponential distribution is not subexponential.

3.6. Suppose that $\xi_1, \ldots, \xi_n$ are independent random variables with common distribution F. Prove the maximum, $\max(\xi_1, \ldots, \xi_n)$, has subexponential distribution if and only if F is subexponential.

3.7. Suppose that F and G are weakly tail-equivalent distributions on $\mathbb{R}^+$. Prove that the convolutions $F * F$ and $G * G$ are weakly tail-equivalent too. Prove the same for n-fold convolutions, for any $n \geq 3$.

3.8. Suppose that $\xi_1, \ldots, \xi_n$ are independent random variables with a common exponential distribution. Find the asymptotics for probability $\mathbb{P}\{\xi_1^{\alpha_1} + \ldots + \xi_n^{\alpha_n} > x\}$ as $x \to \infty$ if

(i) all $\alpha_i > 1$;
(ii) all $\alpha_i < 0$;
(iii) all $\alpha_i \notin [0,1]$.

3.9. Let non-negative random variable ξ have the Weibull distribution with the tail $\overline{F}(x) = e^{-x^\beta}$, $\beta > 0$. For which values of $\alpha > 0$, does $\mathbb{P}\{\xi + \xi^\alpha > x\} \sim \mathbb{P}\{\xi > x\}$ as $x \to \infty$?
Hint: Make use of the equality $\xi + \xi^\alpha = \xi^\alpha(1 + \xi^{1-\alpha})$ and Problem 2.23.

3.10. Let random variable ξ have the standard log-normal distribution. For which values of parameter $\alpha > 0$, does $\mathbb{P}\{\xi + \xi^\alpha > x\} \sim \mathbb{P}\{\xi > x\}$ as $x \to \infty$?

3.11. How do Pitman's criteria work

(i) for the Pareto distribution;
(ii) for the exponential distribution?

3.12. Specify Kesten's bound for the standard Cauchy distribution.

3.13. Let $X(t)$ be a compound Poisson process with a subexponential distribution of jump. For every t, find the asymptotic tail behaviour of the distribution of $X(t)$ in terms of the jump distribution.

3.14. Find an example of subexponential distribution F and of counting random variable τ such that the equivalence

$$\overline{F^{*\tau}}(x) \sim \mathbb{E}\tau \overline{F}(x) \quad \text{as } x \to \infty$$

doesn't hold.
Hint: Make a link to Galton-Watson process.

3.15. *Markov modulated random walk.* Suppose X_n is a time-homogeneous Markov chain with state space $\{1,2\}$ and transition probabilities p_{ij}. Suppose ξ_n, $n \geq 0$, are independent identically distributed random variables with common subexponential distribution F and η_n, $n \geq 0$, are also independent with common subexponential distribution G and such that ξ's and η's are mutually independent. Assume $\zeta_n = \xi_n$ if $X_n = 1$ and $\zeta_n = \eta_n$ if $X_n = 2$. Assume that $\overline{G}(x) \sim c\overline{F}(x)$ as $x \to \infty$, with $0 \leq c < \infty$.

(i) Find the tail asymptotics for the distribution of $\zeta_0 + \zeta_1$ in terms of $\overline{F}$.
(ii) Denote $\tau_1 := \min\{n > 0 : X_n = X_0\}$. Find the tail asymptotics for the distribution of $\zeta_0 + \ldots + \zeta_{\tau_1}$ in terms of $\overline{F}$.
(iii) For $k \geq 2$, let $\tau_k = \min\{n > \tau_{k-1} : X_n = X_0\}$. Find the tail asymptotics for the distribution of $\zeta_0 + \ldots + \zeta_{\tau_k}$ in terms of $\overline{F}$.

3.16. Suppose F is long tailed distribution on $\mathbb{R}^+$ such that the corresponding integrated tail distribution F_I is subexponential. Prove that $(F * F)_I$ is subexponential too.
Hint: Make use of the equality in Problem 2.22.

3.17. Suppose $\xi_1, \ldots, \xi_n$ are independent nonnegative random variables with common subexponential distribution F.

(i) Prove that

$$\mathbb{P}\{\sqrt{\xi_1^2 + \ldots + \xi_n^2} > x\} \sim n\overline{F}(x) \quad \text{as } x \to \infty.$$

(ii) More generally, prove that, for any convex function $g : \mathbb{R}^+ \to \mathbb{R}^+$,

$$\mathbb{P}\{g^{-1}(g(\xi_1) + \ldots + g(\xi_n)) > x\} \sim n\overline{F}(x) \quad \text{as } x \to \infty.$$

3.18. Let random variable ξ have an intermediate regularly varying distribution and random variable η a light-tailed distribution. Prove that, for any joint distribution of ξ and η,

$$\mathbb{P}\{\xi + \eta > x\} \sim \mathbb{P}\{\xi > x\} \quad \text{as } x \to \infty.$$

3.19. Let η be a positive random variable and ξ_1, ξ_2 be two identically distributed random variables which are conditionally independent given any value of η, that is, a.s.

$$\mathbb{P}\{\xi_1 \in B_1, \xi_2 \in B_2 \mid \eta\} = \mathbb{P}\{\xi_1 \in B_1 \mid \eta\}\mathbb{P}\{\xi_2 \in B_2 \mid \eta\}$$

for all Borel sets B_1 and B_2. Find the exact asymptotics for $\mathbb{P}\{\xi_1 > x\}$ and $\mathbb{P}\{\xi_1 + \xi_2 > x\}$ in the following cases:

(i) $\mathbb{P}\{\xi_i > x \mid \eta\} = (1+x)^{-\eta}$ a.s. and η is uniformly distributed in the interval $[1,2]$.

Hint: Make use of the following bounds:

$$\mathbb{P}\{\xi_1 + \xi_2 > x) \geq \mathbb{P}\{\xi_1 > x\} + \mathbb{P}\{\xi_2 > x\} - \mathbb{P}\{\xi_1 > x, \xi_2 > x\}$$

and, for any function $h(x) < x/2$,

$$\mathbb{P}\{\xi_1 + \xi_2 > x\} \leq \mathbb{P}\{\xi_1 > x - h(x)\} + \mathbb{P}\{\xi_2 > x - h(x)\}$$
$$+ \mathbb{P}\{h(x) < \xi_1 < x - h(x), \xi_2 > x - \xi_1\}.$$

(ii) $\mathbb{P}\{\xi_i > x \mid \eta\} = e^{-x^\eta}$ and η is uniformly distributed in the interval $[1/2, 3/2]$.
(iii) $\mathbb{P}\{\xi_i > x \mid \eta\} = e^{-x^\eta}$ and η is uniformly distributed in the interval $[0,1]$.

3.20. Let ξ_1, ξ_2, η_1, η_2 be four mutually independent random variables where ξ_1 and ξ_2 have a regularly varying distribution with parameter $\alpha > 0$ while η_1 and η_2 have a uniform distribution in the interval $[-1,1]$. Find the asymptotics, as $x \to \infty$, for the following probabilities:

(i) $\mathbb{P}\{\xi_1 e^{\eta_1} + \xi_2 e^{\eta_2} > x\};$ \qquad (ii) $\mathbb{P}\{\xi_1 e^{\eta_1} + \xi_2 e^{\eta_1 + \eta_2} > x\}.$

3.21. *Excess process.* In conditions of Problem 2.26, prove that the invariant distribution is subexponential if and only if the integrated tail distribution F_I is subexponential.

3.22. In Problem 2.27, introduce extra conditions that are sufficient for the subexponentiality of the invariant distribution.

Chapter 4
Densities and Local Probabilities

This chapter is devoted to local long-tailedness and to local subexponentiality. First we consider densities with respect to either Lebesgue measure on $\mathbb{R}$ or counting measure on $\mathbb{Z}$. Next we study the asymptotic behaviour of the probabilities to belong to an interval of a fixed length. We give the analogues of the basic properties of the tail probabilities including two analogues of Kesten's estimate, and provide sufficient conditions for probability distributions to have these local properties.

The study of local properties of subexponentiality gives insights into the local asymptotic behaviour of sums and maxima of random variables having heavy-tailed distributions and, in particular, permits us to obtain the local asymptotics for the supremum of a random walk with negative drift. The concept of a subexponential density on the positive line is well-known, while the broader concept of 'delta'-subexponentiality has been introduced recently [4]. The theories for these two classes of distributions look similar, but there are (sometimes essential) differences in the ideas and proofs, and we therefore think that it makes sense to provide a complete treatment of both concepts.

Sections 4.1–4.4 deal with long-tailed densities, subexponential densities, and sufficient conditions for a distribution to have a subexponential density, while Sects. 4.5–4.8 deal with similar topics for Δ-subexponential distributions.

4.1 Long Tailed Densities and Their Convolutions

In this section, we provide the definition and basic properties of long-tailed densities on the real line $\mathbb{R}$. Since a long-tailed density may be a non-monotone function, we cannot prove here a general result similar to Theorem 2.41 for tail distribution functions. We provide instead two separate results, Theorem 4.3 and Lemma 4.4.

Let μ be either Lebesgue measure on $\mathbb{R}$ or counting measure on $\mathbb{Z}$. We say that a distribution F on $\mathbb{R}$ is absolutely continuous with respect to μ if F has a density f with respect to μ, that is, for any Borel set $B \subseteq \mathbb{R}$,

S. Foss et al., *An Introduction to Heavy-Tailed and Subexponential Distributions*,
Springer Series in Operations Research and Financial Engineering,
DOI 10.1007/978-1-4614-7101-1_4, © Springer Science+Business Media New York 2013

$$F(B) = \int_B f(x)\mu(dx).$$

In what follows the argument of the density is either a real number if μ is Lebesgue measure; or an integer if μ is counting measure. If μ is Lebesgue measure, then f is a density of F if, for any Borel set $B \subseteq \mathbb{R}$,

$$F(B) = \int_B f(x)dx.$$

If μ is counting measure, then f is a density of F if, for any $B \subseteq \mathbb{Z}$,

$$F(B) = \sum_{n \in B} f(n).$$

For two distributions F and G with densities f and g respectively, the convolution $F * G$ has density $f * g$ with respect to μ given by

$$(f * g)(x) = \int_{-\infty}^{\infty} f(x - y)G(dy) = \int_{-\infty}^{\infty} f(x - y)g(y)\mu(dy).$$

Definition 4.1. We say that a density f with respect to μ is *long-tailed* if $f(x) > 0$ for all sufficiently large x and $f(x + t) \sim f(x)$ as $x \to \infty$, for any fixed $t > 0$.

Thus a density f is long-tailed if and only if f is a long-tailed function. As pointed out in (2.18), it then follows that $f(x + t) \sim f(x)$ as $x \to \infty$ uniformly over t in compact intervals. In particular, this implies that if f is long-tailed, then $f(x) \to 0$ as $x \to \infty$. To see this, assume that, on the contrary, there exist a sequence $x_n \to \infty$ and $\varepsilon > 0$ such that $x_{n+1} > x_n + 2$ and $f(x_n) \geq 2\varepsilon$ for all n. Then, from the uniform tail-equivalence (2.18) with $a = 1$, there is N such that, for $n \geq N$ and for $x \in [x_n - 1, x_n + 1)$, $f(x) \geq \varepsilon$. Hence,

$$1 = \int_{-\infty}^{\infty} f(y)\mu(dy) \geq \sum_{n=N}^{\infty} \int_{x_n-1}^{x_n+1} f(y)\mu(dy) \geq \sum_{n=N}^{\infty} 2\varepsilon = \infty.$$

This contradiction proves that $f(x) \to 0$ as $x \to \infty$.

Every distribution F with long-tailed density f is long-tailed itself, since for any fixed y

$$\overline{F}(x + y) = \int_0^{\infty} f(x + y + u)\mu(du)$$

$$\sim \int_0^{\infty} f(x + u)\mu(du) = \overline{F}(x) \quad \text{as } x \to \infty.$$

Theorem 4.2. *Let the distributions F and G on $\mathbb{R}$ have densities f and g with respect to μ. Suppose that f is long-tailed. Then the density $f * g$ satisfies*

$$\liminf_{x \to \infty} \frac{(f * g)(x)}{f(x)} \geq 1. \tag{4.1}$$

If, in addition, g is long-tailed, then

$$\liminf_{x \to \infty} \frac{(f * g)(x)}{f(x) + g(x)} = 1. \qquad (4.2)$$

Proof. Fix any $a > 0$. By the uniform convergence (2.18), $f(x - y) \sim f(x)$ as $x \to \infty$ uniformly in $|y| \le a$. Hence,

$$(f * g)(x) \ge \int_{-a}^{a} f(x - y)G(dy) \sim f(x)G[-a, a] \quad \text{as } x \to \infty.$$

Letting $a \to \infty$ we obtain (4.1).

If $g(x)$ is also long-tailed, then $g(x - y) \sim g(x)$ as $x \to \infty$ uniformly in $|y| \le a$. Thus, for all $x > 2a$,

$$(f * g)(x) \ge \int_{-a}^{a} f(x - y)G(dy) + \int_{-a}^{a} g(x - y)F(dy)$$
$$\sim f(x)G[-a, a] + g(x)F[-a, a] \quad \text{as } x \to \infty.$$

Letting $a \to \infty$ we obtain

$$\liminf_{x \to \infty} \frac{(f * g)(x)}{f(x) + g(x)} \ge 1.$$

Hence the equality (4.2) will follow if we show that

$$\liminf_{x \to \infty} \frac{(f * g)(x)}{f(x) + g(x)} \le 1.$$

To prove this, assume that, on the contrary, there exist $\varepsilon > 0$ and x_0 such that, for all $x > x_0$,

$$(f * g)(x) \ge (1 + \varepsilon)(f(x) + g(x)).$$

Integrating with respect to x we obtain

$$\overline{F * G}(x) \ge (1 + \varepsilon)(\overline{F}(x) + \overline{G}(x)),$$

which implies

$$\overline{F^+ * G^+}(x) \ge (1 + \varepsilon)(\overline{F^+}(x) + \overline{G^+}(x)),$$

Since the density f is long-tailed, the distribution F^+ is also long-tailed and, therefore, heavy-tailed, and so the latter inequality contradicts Theorem 2.13. $\square$

Theorem 4.3. *Let the distributions F and G on $\mathbb{R}$ have densities f and g with respect to μ both of which are long-tailed. Then the density $f * g$ of the convolution $F * G$ is also long-tailed.*

Proof. By Lemma 2.19 and Proposition 2.20, we can choose a function h such that $h(x) < x/2$, $h(x) \to \infty$ as $x \to \infty$ and both f and g are h-insensitive (i.e. $f(x-y) \sim f(x)$ and $g(x-y) \sim g(x)$ as $x \to \infty$ uniformly in $|y| \le h(x)$). Fix $t > 0$. Then,

$$(f * g)(x+t) = \int_{-\infty}^{x-h(x)} f(x+t-y)G(dy) + \int_{x-h(x)}^{x+t-h(x)} f(x+t-y)G(dy)$$

$$+ \int_{x+t-h(x)}^{\infty} f(x+t-y)g(y)\mu(dy). \quad (4.3)$$

For fixed $t > 0$, it follows from the given conditions on h that $f(x+t-y) \sim f(x-y)$ as $x \to \infty$ uniformly in $y \le x - h(x)$. Therefore, as $x \to \infty$,

$$\int_{-\infty}^{x-h(x)} f(x+t-y)G(dy) \sim \int_{-\infty}^{x-h(x)} f(x-y)G(dy). \quad (4.4)$$

The second integral is bounded from above by

$$\sup_{y \in [h(x), h(x)+t)} f(y)G[x - h(x), x+t-h(x)) \sim tf(h(x))g(x)$$

$$= o(g(x)) = o((f * g)(x)) \quad (4.5)$$

as $x \to \infty$, by (4.2). The third integral in (4.3) is equal to

$$\int_{-\infty}^{h(x)} g(x+t-y)F(dy) \sim \int_{-\infty}^{h(x)} g(x-y)F(dy) \quad (4.6)$$

by arguments similar to that leading to (4.4). Collecting (4.4)–(4.6), we get $(f * g)(x+t) = (f * g)(x) + o((f * g)(x))$, since the sum of right sides in (4.4) and (4.6) equals $(f * g)(x)$. This completes the proof. $\qquad\qquad\square$

Lemma 4.4. *Let the distributions F and G on $\mathbb{R}$ have densities f and g with respect to μ. Suppose that f is long-tailed and that*

$$\sup_{z \ge x} g(z) = o(f(x)) \quad as \ x \to \infty.$$

Then $f * g$ is also long-tailed.

Proof. Again Lemma 2.19 with Proposition 2.20 enables us to find an increasing function h such that $h(x) < x/2$, $h(x) \to \infty$ as $x \to \infty$ and f is h-insensitive. For any t, consider the following decomposition:

$$(f * g)(x+t) = \int_{-\infty}^{x-h(x)} f(x+t-y)G(dy) + \int_{x-h(x)}^{\infty} f(x+t-y)g(y)\mu(dy).$$

The first integral satisfies (4.4). The second integral is not greater than

$$\sup_{y > x-h(x)} g(y) = o(f(x-h(x))) = o(f(x)). \quad (4.7)$$

It follows from (4.4) and (4.7) that, as $x \to \infty$,

$$(f * g)(x+t) = (1 + o(1))(f * g)(x) + o(f(x)).$$

Applying now the result (4.1) of Theorem 4.2, we arrive at the desired equivalence $(f * g)(x+t) \sim (f * g)(x)$ as $x \to \infty$. $\qquad\square$

Theorems 4.2 and 4.3 imply the following corollary.

Corollary 4.5. *Suppose that f is long-tailed. Then f^{*n} is also long-tailed and*

$$\liminf_{x \to \infty} \frac{f^{*n}(x)}{f(x)} \geq n.$$

4.2 Subexponential Densities on the Positive Half-Line

In this section, we introduce the concept of a subexponential densities on the half real line $\mathbb{R}^+$. We study further their properties, in particular giving closure properties for the class of such densities and providing also the analogue of Kesten's estimate.

Definition 4.6. We say that a density f on $\mathbb{R}^+$ with respect to μ is *subexponential* if f is long-tailed and

$$f^{*2}(x) := \int_0^x f(x-y)f(y)\mu(dy) \sim 2f(x) \quad \text{as } x \to \infty.$$

Typical examples of subexponential densities are given by the Pareto, lognormal, and Weibull (with parameter between 0 and 1) distributions (see Sect. 4.4 for proofs).

Every distribution F with subexponential density f is subexponential itself, since then

$$\overline{F * F}(x) = \int_x^\infty (f * f)(y)\mu(dy)$$

$$\sim 2 \int_x^\infty f(y)\mu(dy)$$

$$= 2\overline{F}(x) \quad \text{as } x \to \infty.$$

The converse result is not in general true: one can, for example, modify a density while keeping the corresponding distribution almost the same. For example, we may take any subexponential density g corresponding to a, necessarily subexponential, distribution G, and construct a new density f such that $f(x)$ is equal to $g(x)$ everywhere except the intervals $x \in [2^n, 2^n+1), n \geq 1$ where we put $f(x) = 0$. To make f a probability density, we may add an appropriate mass to the interval $[0, 2]$. Then the density f is not subexponential because it is not long-tailed. On the other hand, the corresponding distribution F is subexponential, since it may easily be verified that $\overline{F}(x) \sim \overline{G}(x)$ as $x \to \infty$ and G is subexponential.

Now we formulate the basic theorem for subexponential densities.

Theorem 4.7. *Suppose that the distribution F on $\mathbb{R}^+$ has a long-tailed density f with respect to μ. Then the following assertions are equivalent:*

(i) *The density f is subexponential.*

(ii) *For every function h such that $h(x) \to \infty$ as $x \to \infty$ and $h(x) < x/2$,*

$$\int_{h(x)}^{x-h(x)} f(x-y)f(y)\mu(dy) = o(f(x)) \text{ as } x \to \infty. \tag{4.8}$$

(iii) *The relation (4.8) holds for some function h such that $h(x) < x/2$, $h(x) \to \infty$ as $x \to \infty$ and f is h-insensitive.*

Proof. (i)$\Rightarrow$(ii). Assume that f is subexponential. We have

$$f^{*2}(x) = 2\int_0^{h(x)} f(x-y)f(y)\mu(dy) + \int_{h(x)}^{x-h(x)} f(x-y)f(y)dy. \tag{4.9}$$

By Fatou's lemma,

$$\liminf_{x \to \infty} \int_0^{h(x)} \frac{f(x-y)}{f(x)} f(y)\mu(dy) \geq 1,$$

and so (4.8) follows from the subexponentiality of f.

(ii)$\Rightarrow$(iii). This implication is trivial on recalling that f is long-tailed.

(iii)$\Rightarrow$(i). Assume now that the relation (4.8) holds for some function h as given by (iii). Then (4.9) holds, and the first integral on the right of (4.9) is tail-equivalent to $f(x)$ (as $x \to \infty$) by the choice of the function h. Together with the condition (4.8) this implies the subexponentiality of f. □

Theorem 4.8. *Let f be a subexponential density on $\mathbb{R}^+$ with respect to μ. Suppose that the density g on $\mathbb{R}^+$ is long-tailed and that f and g are weakly tail-equivalent, that is,*

$$0 < \liminf_{x \to \infty} \frac{g(x)}{f(x)} \leq \limsup_{x \to \infty} \frac{g(x)}{f(x)} < \infty. \tag{4.10}$$

Then g is also subexponential.

In particular, the condition (4.10) is satisfied if $g(x) \sim cf(x)$ as $x \to \infty$ for some $c \in (0, \infty)$.

Proof. The result follows from Theorem 4.7(ii) and (iii): observe that (4.10) implies that there exists $c_1 < \infty$ such that $g(x) \leq c_1 f(x)$ for all sufficiently large x; hence, for any function h such that $h(x) < x/2$ for all x, $h(x) \to \infty$ as $x \to \infty$ and g is h-insensitive,

$$\int_{h(x)}^{x-h(x)} g(x-y)g(y)\mu(dy) \le c_1^2 \int_{h(x)}^{x-h(x)} f(x-y)f(y)\mu(dy) = o(f(x)) = o(g(x)).$$

$\square$

Lemma 4.9. *Let f be a subexponential density on $\mathbb{R}^+$ with respect to μ. Let f_1, f_2 be two densities on $\mathbb{R}^+$ such that $f_1(x)/f(x) \to c_1$ and $f_2(x)/f(x) \to c_2$ as $x \to \infty$, for some constants c_1, $c_2 \ge 0$. Then*

$$\frac{(f_1 * f_2)(x)}{f(x)} \to c_1 + c_2 \quad as \; x \to \infty. \tag{4.11}$$

Further, if $c_1 + c_2 > 0$ then the convolution $f_1 * f_2$ is a subexponential density.

Proof. Let h be any function such that $h(x) < x/2$, $h(x) \to \infty$ as $x \to \infty$ and f is h-insensitive. Then

$$f_1 * f_2(x) = \int_0^{h(x)} f_1(x-y)f_2(y)\mu(dy) + \int_0^{h(x)} f_2(x-y)f_1(y)\mu(dy)$$
$$+ \int_{h(x)}^{x-h(x)} f_1(x-y)f_2(y)\mu(dy)$$
$$=: I_1(x) + I_2(x) + I_3(x).$$

We have $I_1(x)/f(x) \to c_1$ and $I_2(x)/f(x) \to c_2$ as $x \to \infty$. Finally,

$$I_3(x) \le (c_1 c_2 + o(1)) \int_{h(x)}^{x-h(x)} f(x-y)f(y)\mu(dy) = o(f(x)),$$

by Theorem 4.7(ii), so that (4.11) now follows. The final assertion of the lemma follows from Theorem 4.8. $\square$

Using induction arguments, we obtain the following corollary.

Corollary 4.10. *Assume that f is a subexponential density on $\mathbb{R}^+$ with respect to μ. Then, for any $n \ge 2$, $f^{*n}(x) \sim nf(x)$ as $x \to \infty$ and f^{*n} is a subexponential density.*

For subexponential densities on $\mathbb{R}^+$ we have the following analogue of Kesten's bound.

Theorem 4.11. *Assume that f is a subexponential density on $\mathbb{R}^+$ with respect to μ. If f is bounded, then, for any $\varepsilon > 0$, there exist $x_0 = x_0(\varepsilon)$ and $c(\varepsilon) > 0$ such that, for any $x > x_0$ and for any integer $n \ge 1$,*

$$f^{*n}(x) \le c(\varepsilon)(1+\varepsilon)^n f(x).$$

Proof. Take $c < \infty$ such that $f(x) \le c$ for all $x \ge 0$. Then it follows from the convolution formula that

$$f^{*n}(x) \le cF^{*(n-1)}[0,x] \le c \quad \text{for all } x \ge 0 \text{ and } n \ge 1. \tag{4.12}$$

Since f is long-tailed, there exists x_1 such that

$$\inf_{x \in [x_1, x_2]} f(x) > 0 \quad \text{for every } x_2 > x_1. \tag{4.13}$$

For $x_0 > x_1$ and $n \geq 1$, put

$$A_n(x_0) := \sup_{x > x_0} \frac{f^{*n}(x)}{f(x)}.$$

Fix any $\varepsilon > 0$. By the subexponentiality of f, there exists x_0 such that, for all $x > x_0$,

$$\int_0^{x-x_0} f(x-y)f(y)\mu(dy) \leq (1+\varepsilon/2)f(x).$$

For any $n \geq 2$ and $x > 2x_0$,

$$f^{*n}(x) = \int_0^{x-x_0} f^{*(n-1)}(x-y)f(y)\mu(dy) + \int_0^{x_0} f(x-y)f^{*(n-1)}(y)\mu(dy).$$

By the definition of $A_{n-1}(x_0)$ and the choice of x_0,

$$\int_0^{x-x_0} f^{*(n-1)}(x-y)f(y)\mu(dy) \leq A_{n-1}(x_0) \int_0^{x-x_0} f(x-y)f(y)\mu(dy)$$
$$\leq A_{n-1}(x_0)(1+\varepsilon/2)f(x). \tag{4.14}$$

Further,

$$\int_0^{x_0} f(x-y)f^{*(n-1)}(y)\mu(dy) \leq \max_{0 < y \leq x_0} f(x-y) \leq L_1 f(x), \tag{4.15}$$

where

$$L_1 := \sup_{0 < y \leq x_0, t > 2x_0} \frac{f(t-y)}{f(t)}.$$

If $x_0 < x \leq 2x_0$, then, by (4.12) and (4.13),

$$\frac{f^{*n}(x)}{f(x)} \leq \frac{c}{\inf_{x_0 < t \leq 2x_0} f(t)} =: L_2 < \infty. \tag{4.16}$$

Since f is long-tailed, we may choose x_0 so that also $L_1 < \infty$. Put $L = \max(L_1, L_2)$. It follows from (4.14) to (4.16) that, for any $x > x_0$,

$$f^{*n}(x) \leq (A_{n-1}(x_0)(1+\varepsilon/2)+L)f(x).$$

Hence, $A_n(x_0) \leq A_{n-1}(x_0)(1+\varepsilon/2)+L$. Therefore, an induction argument gives

$$A_n(x_0) \leq A_1(x_0)(1+\varepsilon/2)^{n-1}+L\sum_{l=0}^{n-2}(1+\varepsilon/2)^l \leq Ln(1+\varepsilon/2)^{n-1},$$

which implies the conclusion of the theorem.   □

4.3 Subexponential Densities on the Real Line

In the previous chapter we showed that there are two equivalent ways to define the subexponentiality of probability distributions on the whole real line: the distribution F of a random variable ξ is subexponential if, either:

(1) It is long-tailed and $\overline{F * F}(x) \sim 2\overline{F}(x)$ as $x \to \infty$, or, equivalently, if
(2) The conditional distribution $\mathbb{P}\{\xi \in \cdot \,|\, \xi \geq 0)$ on the positive half-line $\mathbb{R}^+$ is subexponential

The latter shows that subexponentiality continues to be a tail property for distributions defined on the whole real line; in particular we have Theorem 3.6.

For probability densities the situation is more complex. Suppose that a distribution F on $\mathbb{R}$ has a density f with respect to μ. We give here two candidate conditions for the definition of the subexponentiality of f, the first of which preserves the tail property while the second does not. In the interests of openness we refrain from making a present judgement as to which condition might be considered more appropriate. Our two conditions are:

(D1) The density f^+ of the conditional distribution of F on $\mathbb{R}^+$, that is,

$$f^+(x) := \frac{f(x)\mathbb{I}\{x \geq 0\}}{F(\mathbb{R}^+)}, \qquad (4.17)$$

is subexponential.

(D2) The density f of F is long-tailed and

$$f^{*2}(x) := \int_{-\infty}^{\infty} f(x-y)f(y)\mu(dy) \sim 2f(x) \quad \text{as } x \to \infty. \qquad (4.18)$$

It follows immediately from the results of the previous section that the condition (D1) defines a tail property of the distribution F. Further, if the distribution F has a support which is bounded below, i.e. $F(a) = 0$ for some $a \in \mathbb{R}$, then the situation is essentially no different from that in which F in concentrated on $\mathbb{R}^+$; it is easy to see (using the long-tailedness of f) that the theory of the previous section continues to hold and that the conditions (D1) and (D2) above are equivalent. However, when the support of the distribution F is not bounded below, then the behaviour of its left tail may influence the right tail of the density of its convolution with itself, and here the condition (D2) may not correspond to a tail property of F.

In Lemma 4.12 below we show that the condition (D2) implies the condition (D1), while in Lemma 4.13 we show that, under a further condition on the *right* tail of F, the conditions (D1) and (D2) are equivalent.

Lemma 4.12. *Suppose that the distribution F on $\mathbb{R}$ has a density f with respect to μ which is long-tailed and which satisfies the condition (4.18). Then the density f^+ on $\mathbb{R}^+$ given by (4.17) is subexponential.*

Proof. Since f is long-tailed, by Lemma 2.19, we can choose a function h such that $h(x) < x/2$, $h(x) \to \infty$ as $x \to \infty$ and f is h-insensitive. Then

$$f^{*2}(x) = 2 \int_{-\infty}^{-h(x)} f(x-y)f(y)\mu(dy) + 2 \int_{-h(x)}^{h(x)} f(x-y)f(y)\mu(dy)$$

$$+ \int_{h(x)}^{x-h(x)} f(x-y)f(y)\mu(dy) \tag{4.19}$$

$$\geq 2 \int_{-h(x)}^{h(x)} f(x-y)f(y)\mu(dy) + \int_{h(x)}^{x-h(x)} f(x-y)f(y)\mu(dy)$$

$$\sim 2f(x) + \int_{h(x)}^{x-h(x)} f(x-y)f(y)\mu(dy)$$

as $x \to \infty$, by the choice of the function h. Since also f satisfies the condition (4.18), we obtain that

$$\int_{h(x)}^{x-h(x)} f(x-y)f(y)\mu(dy) = o(f(x)) \quad \text{as } x \to \infty.$$

Hence the density f^+ satisfies the condition (4.8) of Theorem 4.7 and so is subexponential. $\qquad\square$

We now give the converse result where we require an extra condition.

Lemma 4.13. *Suppose that the distribution F on $\mathbb{R}$ has a density f with respect to μ such that, for some x_0 and $c < \infty$,*

$$f(x+y) \leq cf(x) \quad \text{for all } x > x_0 \text{ and } y > 0. \tag{4.20}$$

Suppose further that the density f^+ defined by (4.17) is subexponential. Then f (in addition to being long-tailed) satisfies the condition (4.18).

Proof. Since the subexponentiality of f^+ implies that f is long-tailed, we may again choose a function h such that $h(x) < x/2$, $h(x) \to \infty$ as $x \to \infty$ and f is h-insensitive. We make use of decomposition (4.19). It follows from the condition (4.20) that

$$\int_{-\infty}^{-h(x)} f(x-y)f(y)\mu(dy) \leq cf(x)F(-h(x))$$

$$= o(f(x)) \quad \text{as } x \to \infty. \tag{4.21}$$

Further, from the choice of h,

$$\int_{-h(x)}^{h(x)} f(x-y)f(y)\mu(dy) \sim f(x) \int_{-h(x)}^{h(x)} f(y)\mu(dy)$$

$$\sim 2f(x) \quad \text{as } x \to \infty. \tag{4.22}$$

Since the density f^+ is subexponential, it follows from Theorem 4.7 that

$$\int_{h(x)}^{x-h(x)} f^+(x-y)f^+(y)\mu(dy) = o(f(x)) \quad \text{as } x \to \infty,$$

and hence

$$\int_{h(x)}^{x-h(x)} f(x-y)f(y)\mu(dy) = o(f(x)) \quad \text{as } x \to \infty. \tag{4.23}$$

The relations (4.21), (4.22), and (4.23) now imply that $f^{*2}(x) \sim 2f(x)$ as $x \to \infty$.

$\square$

We conclude this section with the following comment. In contrast to tail functions, densities are not in general decreasing functions. Moreover, a subexponential density may be not tail-equivalent to any non-increasing function and, in particular, the condition (4.20) may fail. This may be viewed as being essentially the reason for the difficulty in extending the concept of subexponentiality of densities to distributions on $\mathbb{R}$.

4.4 Sufficient Conditions for Subexponentiality of Densities

Sufficient conditions for distributions to be subexponential were given in Sect. 3.5. In this section, we provide similar conditions for subexponentiality of densities.

Theorem 4.14. *Let the distribution F on $\mathbb{R}^+$ have a long-tailed density f. Suppose that there exist $c > 0$ and x_0 such that $f(y) \geq cf(x)$ for any $x > x_0$ and $y \in (x, 2x]$. Then the density f is subexponential.*

Proof. Let h be any positive function such that $h(x) \to \infty$ as $x \to \infty$ and $h(x) < x/2$ for all x. Then

$$\int_{h(x)}^{x-h(x)} f(x-y)f(y)\mu(dy) = 2\int_{h(x)}^{x/2} f(x-y)f(y)\mu(dy)$$

$$\leq 2cf(x)\int_{h(x)}^{x/2} f(y)\mu(dy) = o(f(x))$$

as $x \to \infty$. The subexponentiality of f now follows from Theorem 4.7(ii). $\square$

Observe that in particular the density of the Pareto distribution satisfies the conditions of Theorem 4.14.

Theorem 4.15. *Let the distribution F on $\mathbb{R}^+$ have a long-tailed density f. Suppose that, for some x_0, the function $R(x) := -\ln f(x)$ is concave for $x \geq x_0$. Suppose further that there exists a function h such that $h(x) < x/2$ for all x, that $h(x) \to \infty$ as $x \to \infty$, that f is h-insensitive, and that $xe^{-R(h(x))} \to 0$ as $x \to \infty$. Then the density f is subexponential.*

Proof. Since g is concave, the minimum of the sum $R(x-y)+R(y)$ in $y \in [h(x), x-h(x)]$ is equal to $R(x-h(x))+R(h(x))$. Therefore,

$$\int_{h(x)}^{x-h(x)} f(x-y)f(y)\mu(dy) = \int_{h(x)}^{x-h(x)} e^{-(R(x-y)+R(y))}\mu(dy) \leq xe^{-(R(x-h(x))+R(h(x)))}.$$

Since $e^{-R(x-h(x))} \sim e^{-R(x)}$,

$$\int_{h(x)}^{x-h(x)} f(x-y)f(y)\mu(dy) = O(e^{-R(x)}xe^{-R(h(x))}) = o(f(x)),$$

so that the result now follows from Theorem 4.7.   □

The density of the Weibull distribution with parameter $\alpha \in (0,1)$ satisfies conditions of Theorem 4.15 with $h(x) = \ln^{2/\alpha} x$. The density of the lognormal distribution satisfies the these conditions with $h(x) = \sqrt{x}$.

4.5 Δ-Long-Tailed Distributions and Their Convolutions

This section and the next deal with local properties of long-tailedness and subexponentiality which may be considered as intermediate properties of a distribution between that of being long-tailed/subexponential and that of having a long-tailed/subexponential density, and are formulated in terms of the probability for a random variable to belong to an interval of a fixed length when the location of the interval is tending to infinity.

Define $\Delta = (0,T]$ for some finite $T > 0$. For any x and for any nonnegative integer n, define also $x+\Delta := (x,x+T]$ and $n\Delta := (0,nT]$.

We now introduce the following definition.

Definition 4.16. A distribution F on $\mathbb{R}$ is called Δ-*long-tailed* if $F(x+\Delta)$ is a long-tailed function, that is, for any fixed $y > 0$,

$$\frac{F(x+y+\Delta)}{F(x+\Delta)} \to 1 \quad \text{as } x \to \infty.$$

By the property (2.18) of long-tailed functions, the latter convergence holds uniformly over all y in any compact set. We write also $\mathcal{L}_\Delta$ for the class of Δ-long-tailed distributions. We consider here only finite intervals Δ, but if we allowed the interval to be infinite, $\Delta = (0,\infty)$, we would have $\mathcal{L}_\Delta = \mathcal{L}$, the class of long-tailed distributions.

It follows from the definition that, if $F \in \mathcal{L}_\Delta$ for some interval $\Delta = (0,T]$, then $F \in \mathcal{L}_{n\Delta}$ for any $n = 2,3,\dots$ and also $F \in \mathcal{L}$. To see this observe that, for any fixed $y > 0$ and any $n \in \{2,3,\dots,\infty\}$,

$$F(x+y+n\Delta) = \sum_{k=0}^{n-1} F(x+kT+y+\Delta)$$

$$\sim \sum_{k=0}^{n-1} F(x+kT+\Delta) = F(x+n\Delta).$$

Note that any distribution F on the integer lattice with $F\{n+1\} \sim F\{n\}$ as $n \to \infty$ (i.e. with a long-tailed density with respect to counting measure) may be also viewed as a member of $\mathcal{L}_\Delta$ with $\Delta = (0,1]$.

In earlier chapters we dealt with tail functions of distributions (for any distribution F and for any x, the tail $\overline{F}(x) = F(x+\Delta)$ with $\Delta = (0,\infty)$). Tail functions are monotone non-increasing, and this allowed us to prove Theorem 2.11. For finite intervals Δ, there is in general no such monotonicity, and we need further restrictions, given by Theorem 4.17, to obtain the inequality

$$\liminf_{x\to\infty} \frac{(F*G)(x+\Delta)}{F(x+\Delta)+G(x+\Delta)} \geq 1. \tag{4.24}$$

Theorem 4.17. *Let the distributions F and G belong to the class $\mathcal{L}_\Delta$, where $\Delta = (0,T]$ for some finite T. Then*

$$\liminf_{x\to\infty} \frac{(F*G)(x+\Delta)}{F(x+\Delta)+G(x+\Delta)} = 1. \tag{4.25}$$

Proof. Let ξ and η be two independent random variables with respective distributions F and G. Fix any $a > 0$. For $x > 2a$, we have the following lower bound:

$$(F*G)(x+\Delta) \geq \mathbb{P}\{\xi+\eta \in x+\Delta, \, |\xi| \leq a\} + \mathbb{P}\{\xi+\eta \in x+\Delta, \, |\eta| \leq a\}.$$

We also have the tail equivalences $F(x+y+\Delta) \sim F(x+\Delta)$ and $G(x+y+\Delta) \sim G(x+\Delta)$ as $x \to \infty$ uniformly in $|y| \leq a$. Therefore, as $x \to \infty$,

$$\mathbb{P}\{\xi+\eta \in x+\Delta, \, |\xi| \leq a\} = \int_{[-a,a]} G(x-y+\Delta)F(dy)$$

$$\sim G(x+\Delta) \int_{[-a,a]} F(dy)$$

$$\sim G(x+\Delta)F[-a,a],$$

and similarly

$$\mathbb{P}\{\xi+\eta \in x+\Delta, \, |\eta| \leq a\} \sim F(x+\Delta)G[-a,a].$$

Letting $a \to \infty$ implies the lower bound (4.24). Now assume that, on the contrary, the equality (4.25) does not hold, that is,

$$\liminf_{x\to\infty} \frac{(F*G)(x+\Delta)}{F(x+\Delta)+G(x+\Delta)} > 1.$$

Then there exist $\varepsilon > 0$ and x_0 such that, for all $x > x_0$ and $n \geq 0$,

$$(F * G)(x + nT + \Delta) \geq (1 + \varepsilon)(F(x + nT + \Delta) + G(x + nT + \Delta)).$$

Summing over $n \geq 0$, we obtain

$$\overline{F * G}(x) \geq (1 + \varepsilon)(\overline{F}(x) + \overline{G}(x)),$$

which implies

$$\overline{F^+ * G^+}(x) \geq (1 + \varepsilon)(\overline{F^+}(x) + \overline{G^+}(x)),$$

However, since $F \in \mathcal{L}_\Delta \subseteq \mathcal{L}$, it follows that the distribution F^+ is heavy-tailed, and therefore the latter inequality contradicts Theorem 2.13. $\square$

In the next theorem we prove that, for any Δ, the class $\mathcal{L}_\Delta$ is closed under convolutions.

Theorem 4.18. *Let the distributions F and G belong to the class $\mathcal{L}_\Delta$ for some finite interval $\Delta = (0, T]$. Then $F * G \in \mathcal{L}_\Delta$.*

Proof. Let ξ and η be two independent random variables with respective distributions F and G. By Lemma 2.19 and Proposition 2.20 there exists a function h such that $h(x) < x/2$, $h(x) \to \infty$ and both $F(x + \Delta)$ and $G(x + \Delta)$ are h-insensitive.

Consider the event $B(x, t) = \{\xi + \eta \in x + t + \Delta\}$. In order to prove that $F * G \in \mathcal{L}_\Delta$, we need to check that, for any $t > 0$, $\mathbb{P}\{B(x, t)\} \sim \mathbb{P}\{B(x, 0)\}$ as $x \to \infty$. Since the events $\{\xi \leq x - h(x)\}$ and $\{\eta \leq h(x)\}$ together imply $\{\xi + \eta \leq x\}$, we have the following decomposition:

$$\mathbb{P}\{B(x, t)\} = \mathbb{P}\{B(x, t), \, \xi \leq x - h(x)\}$$

$$+ \mathbb{P}\{B(x, t), \, \eta \leq h(x)\} + \mathbb{P}\{B(x, t), \, \xi > x - h(x), \, \eta > h(x)\}. \quad (4.26)$$

For fixed $t > 0$, $G(x + t - y + \Delta) \sim G(x - y + \Delta)$ as $x \to \infty$ uniformly in $y \leq x - h(x)$, since $h(x) \to \infty$. Therefore, as $x \to \infty$,

$$\mathbb{P}\{B(x, t), \, \xi \leq x - h(x)\} = \int_{-\infty}^{x - h(x)} G(x + t - y + \Delta) F(dy)$$

$$\sim \int_{-\infty}^{x - h(x)} G(x - y + \Delta) F(dy)$$

$$= \mathbb{P}\{B(x, 0), \, \xi \leq x - h(x)\}. \quad (4.27)$$

A similar argument shows that

$$\mathbb{P}\{B(x,t),\ \eta \leq h(x)\} = \int_{-\infty}^{h(x)} F(x+t-y+\Delta)G(dy)$$

$$\sim \int_{-\infty}^{h(x)} F(x-y+\Delta)G(dy)$$

$$= \mathbb{P}\{B(x,0),\ \eta \leq h(x)\}. \tag{4.28}$$

Finally,

$$\mathbb{P}\{B(x,t),\ \xi > x - h(x),\ \eta > h(x)\}$$
$$= \mathbb{P}\{B(x,t),\ \xi \in (x-h(x), x-h(x)+t+T],\ \eta > h(x)\}.$$

The value of the latter probability is at most

$$\overline{G}(h(x))F(x-h(x)+(0,t+T]) = o(F(x-h(x)+(0,t+T])) \quad \text{as } x \to \infty.$$

Without loss of generality, we can assume that $t < T$. Then,

$$F(x-h(x)+(0,t+T]) \leq F(x-h(x)+\Delta) + F(x-h(x)+T+\Delta).$$

Both terms on the right side of the above expression are of the order $O(F(x+\Delta))$, by the choice of the function h. Thus, as $x \to \infty$,

$$\mathbb{P}\{B(x,t),\ \xi > x - h(x),\ \eta > h(x)\} = o(F(x+\Delta)). \tag{4.29}$$

Combining (4.26)–(4.29) we conclude that

$$\mathbb{P}\{B(x,t)\} = (1+o(1))\mathbb{P}\{B(x,0)\} + o(F(x+\Delta))$$

as $x \to \infty$. The conclusion of the theorem now follows on applying Theorem 4.17. $\quad\square$

By induction arguments we obtain the following corollary to Theorems 4.17 and 4.18.

Corollary 4.19. *If $F \in \mathcal{L}_\Delta$, then, for all $n \geq 2$, $F^{*n} \in \mathcal{L}_\Delta$ and*

$$\liminf_{x \to \infty} \frac{F^{*n}(x+\Delta)}{F(x+\Delta)} \geq n.$$

4.6 Δ-Subexponential Distributions

We continue our study of local properties by introducing the concept of Δ-subexponentiality.

Definition 4.20. Let F be a distribution on $\mathbb{R}^+$ with right-unbounded support. For any fixed $\Delta = (0, T]$ for some finite $T > 0$ we say that F is Δ-*subexponential* if $F \in \mathcal{L}_\Delta$ and

$$(F * F)(x + \Delta) \sim 2F(x + \Delta) \quad \text{as } x \to \infty.$$

Equivalently, a random variable ξ has a Δ-subexponential distribution if the function $\mathbb{P}\{\xi \in x + \Delta\}$ is long-tailed and, for two independent copies ξ_1 and ξ_2 of ξ,

$$\mathbb{P}\{\xi_1 + \xi_2 \in x + \Delta\} \sim 2\mathbb{P}\{\xi \in x + \Delta\} \quad \text{as } x \to \infty.$$

In this and the following two sections, we always consider finite intervals Δ. But if we allowed the interval to be infinite, $\Delta = (0, \infty)$, then the class of $(0, \infty)$-subexponential distributions would be none other than the standard class $\mathcal{S}$ of subexponential distributions on $\mathbb{R}^+$. For all finite Δ, the typical examples of Δ-subexponential distributions are the same—in particular the Pareto, lognormal, and Weibull (with parameter between 0 and 1) distributions, as we shall show in Sect. 4.8. Also, many properties of Δ-subexponential distributions with finite Δ are very close to those of subexponential distributions, as we shall show below. However, we have to repeat (see the previous section) that, for any distribution F, in contrast to the tail function $\overline{F}$, the function $F(x + \Delta)$ may be non-monotone. This leads to extra challenges in the study of Δ-subexponentiality (see the example on non-monotonicity at the end of Sect. 4.8).

Note that, for any $\Delta = (0, T)$, any distribution F with subexponential density f is Δ-subexponential since

$$(F * F)(x + \Delta) = \int_x^{x+T} (f * f)(y)\mu(dy)$$

$$\sim 2 \int_x^{x+T} f(y)\mu(dy) = 2F(x + \Delta) \quad \text{as } x \to \infty. \quad (4.30)$$

Further, it follows from the definition that, if F is Δ-subexponential, then F is $n\Delta$-subexponential for any $n = 2, 3, \ldots$ and $F \in \mathcal{S}_\mathbb{R}$. To see this observe that, for any $n \in \{2, 3, \ldots, \infty\}$ and as $x \to \infty$,

$$\mathbb{P}\{\xi_1 + \xi_2 \in x + n\Delta\} = \sum_{k=0}^{n-1} \mathbb{P}\{\xi_1 + \xi_2 \in x + kT + \Delta\}$$

$$\sim 2 \sum_{k=0}^{n-1} \mathbb{P}\{\xi \in x + kT + \Delta\}$$

$$= 2\mathbb{P}\{\xi \in x + n\Delta\}.$$

Thus we have in particular that, for any Δ, the class of Δ-subexponential distributions is a subclass of $\mathcal{S}$.

Note also that if we consider the distributions concentrated on the integers, then the class of $(0, 1]$-subexponential distributions consists of all distributions F such that $F\{n + 1\} \sim F\{n\}$ and $F^{*2}\{n\} \sim 2F\{n\}$ as $n \to \infty$, and so coincides with the class of distributions with subexponential densities.

We now have the following theorem which characterises Δ-subexponential distributions on the positive half-line $\mathbb{R}^+$ and which is analogous to Theorem 3.6 or Theorem 3.7 for subexponential distributions on $\mathbb{R}$ and to Theorem 4.7 for subexponential densities on $\mathbb{R}^+$.

Theorem 4.21. *Suppose that the distribution F on $\mathbb{R}^+$ is such that $F \in \mathcal{L}_\Delta$ for some Δ. Let ξ_1 and ξ_2 be two independent random variables with common distribution F. Then the following assertions are equivalent:*
 (i) *F is Δ-subexponential.*
 (ii) *For every function h such that $h(x) \to \infty$ as $x \to \infty$ and $h(x) < x/2$,*

$$\mathbb{P}\{\xi_1 + \xi_2 \in x + \Delta, \xi_1 > h(x), \xi_2 > h(x)\} = o(F(x + \Delta)). \qquad (4.31)$$

(iii) *There exists a function h such that $h(x) < x/2$, $h(x) \to \infty$ as $x \to \infty$, the function $F(x + \Delta)$ is h-insensitive and the relation (4.31) holds.*

Proof. (i)$\Rightarrow$(ii). Suppose first that F is Δ-subexponential. Define the event $B = \{\xi_1 + \xi_2 \in x + \Delta\}$. We have

$$\mathbb{P}\{B\} = \mathbb{P}\{B, \xi_1 \le h(x)\} + \mathbb{P}\{B, \xi_2 \le h(x)\} + \mathbb{P}\{B, \xi_1 > h(x), \xi_2 > h(x)\}$$
$$= 2\mathbb{P}\{B, \xi_1 \le h(x)\} + \mathbb{P}\{B, \xi_1 > h(x), \xi_2 > h(x)\}. \qquad (4.32)$$

By Fatou's lemma,

$$\liminf_{x \to \infty} \frac{\mathbb{P}\{B, \xi_1 \le h(x)\}}{F(x + \Delta)} = \liminf_{x \to \infty} \int_0^{h(x)} \frac{F(x - y + \Delta)}{F(x + \Delta)} F(dy) \ge 1. \qquad (4.33)$$

Hence from (4.32), (4.33) and the Definition 4.20 of Δ-subexponentiality, we obtain (4.31).

That (ii) implies (iii) is trivial since the condition $F \in \mathcal{L}_\Delta$ implies the existence of a function with respect to which $F(x + \Delta)$ is h-insensitive.

(iii)$\Rightarrow$(i). Now suppose that the condition (iii) holds for some function h. We again use the decomposition (4.32) for B as defined above. Then

$$\mathbb{P}\{B, \xi_1 \le h(x)\} = \int_0^{h(x)} F(x - y + \Delta)F(dy)$$
$$\sim F(x + \Delta) \int_0^{h(x)} F(dy)$$
$$\sim F(x + \Delta) \qquad \text{as } x \to \infty,$$

and so (4.31) together with (4.32) implies the Δ-subexponentiality of F. $\qquad\square$

Next we prove the result which shows in particular that the subclass of Δ-subexponential distributions on $\mathbb{R}^+$ is closed under the natural Δ-equivalence relation.

Theorem 4.22. *Let F be a Δ-subexponential distribution on $\mathbb{R}^+$, for some Δ. Suppose that the distribution G on $\mathbb{R}^+$ belongs to $\mathcal{L}_\Delta$ and the functions $F(x+\Delta)$ and $G(x+\Delta)$ are weakly tail-equivalent, that is,*

$$0 < \liminf_{x\to\infty} \frac{G(x+\Delta)}{F(x+\Delta)} \le \limsup_{x\to\infty} \frac{G(x+\Delta)}{F(x+\Delta)} < \infty. \tag{4.34}$$

Then G is also Δ-subexponential. In particular, G is Δ-subexponential provided $G(x+\Delta) \sim cF(x+\Delta)$ as $x \to \infty$ for some $c > 0$.

Proof. Choose a function h such that $h(x) < x/2$ for all x, $h(x) \to \infty$ as $x \to \infty$ and the function $G(x+\Delta)$ is h-insensitive. Let $\xi_1, \xi_2, \zeta_1, \zeta_2$ be independent random variables such that ξ_1 and ξ_2 have common distribution F, and ζ_1 and ζ_2 have common distribution G. By Theorem 4.21, it is sufficient to prove that

$$\mathbb{P}\{\zeta_1 + \zeta_2 \in x+\Delta, \zeta_1 > h(x), \zeta_2 > h(x)\} = o(G(x+\Delta)).$$

The probability on the left side of the above expression is not greater than

$$\int_{h(x)}^{x-h(x)+T} G(x-y+\Delta)G(dy) =: I.$$

By the condition (4.34), for some $c_1 < \infty$ and for all sufficiently large x,

$$I \le c_1 \int_{h(x)}^{x-h(x)+T} F(x-y+\Delta)G(dy)$$
$$\le c_1 \mathbb{P}\{\zeta_1 + \xi_2 \in x+\Delta, \zeta_1 > h(x), \xi_2 > h(x) - T\}$$
$$\le c_1 \int_{h(x)-T}^{x-h(x)+2T} G(x-y+\Delta)F(dy).$$

A repetition of the above argument now gives that

$$I \le c_1^2 \int_{h(x)-T}^{x-h(x)+2T} F(x-y+\Delta)F(dy)$$
$$\le c_1^2 \mathbb{P}\{\xi_1 + \xi_2 \in x+\Delta, \xi_1 \ge h(x) - 2T, \xi_2 \ge h(x) - T\}$$
$$= o(F(x+\Delta))$$
$$= o(G(x+\Delta)).$$

as required, where the third line in the above display again follows from Theorem 4.21. $\qquad\square$

Theorem 4.23. *Suppose that the distribution F on $\mathbb{R}^+$ is Δ-subexponential, for some Δ. Let G_1, G_2 be two distributions on $\mathbb{R}^+$ such that $G_1(x+\Delta)/F(x+\Delta) \to c_1$ and $G_2(x+\Delta)/F(x+\Delta) \to c_2$ as $x \to \infty$, for some constants $c_1, c_2 \ge 0$. Then*

$$\frac{(G_1 * G_2)(x+\Delta)}{F(x+\Delta)} \to c_1 + c_2 \quad as\ x \to \infty. \tag{4.35}$$

Further, if $c_1 + c_2 > 0$ then the convolution $G_1 * G_2$ is Δ-subexponential.

Proof. Let ζ_1 and ζ_2 be independent random variables with distributions G_1 and G_2 respectively. Let h be a function such that $h(x) < x/2$ for all x, $h(x) \to \infty$ as $x \to \infty$, and the function $F(x+\Delta)$ is h-insensitive. Define also the event $B = \{\zeta_1 + \zeta_2 \in x+\Delta\}$. Then

$$\mathbb{P}\{B\} = \mathbb{P}\{B, \zeta_1 \leq h(x)\} + \mathbb{P}\{B, \zeta_2 \leq h(x)\} + \mathbb{P}\{B, \zeta_1 > h(x), \zeta_2 > h(x)\}. \quad (4.36)$$

As in the last lines of the proof of Theorem 4.21, one can show that

$$\mathbb{P}\{B, \zeta_1 \leq h(x)\} \sim G_2(x+\Delta), \quad \mathbb{P}\{B, \zeta_2 \leq h(x)\} \sim G_1(x+\Delta)$$

as $x \to \infty$, and then

$$\frac{\mathbb{P}\{B, \zeta_1 \leq h(x)\}}{F(x+\Delta)} \to c_2, \qquad \frac{\mathbb{P}\{B, \zeta_2 \leq h(x)\}}{F(x+\Delta)} \to c_1. \quad (4.37)$$

Following the same argument as that in the proof of Theorem 4.22, we obtain also that

$$\mathbb{P}\{B, \zeta_1 > h(x), \zeta_2 > h(x)\} = o(F(x+\Delta)). \quad (4.38)$$

The result (4.35) now follows from (4.36) to (4.38).

The final assertion of the theorem follows from Theorem 4.22. $\quad\square$

By induction, Theorem 4.23 implies the following corollary.

Corollary 4.24. *Suppose that the distribution F on $\mathbb{R}^+$ is Δ-subexponential, for some Δ. Let G be a distribution on $\mathbb{R}^+$ such that $G(x+\Delta)/F(x+\Delta) \to c \geq 0$ as $x \to \infty$. Then, for any $n \geq 2$, $G^{*n}(x+\Delta)/F(x+\Delta) \to nc$ as $x \to \infty$. If $c > 0$, then G^{*n} is Δ-subexponential.*

We conclude this section with the result which provides of Kesten's upper bound for the class of Δ-subexponential distributions.

Theorem 4.25. *Suppose that the distribution F on $\mathbb{R}^+$ is Δ-subexponential, for some $\Delta = (0, T]$. Then, for any $\varepsilon > 0$, there exist $x_0 = x_0(\varepsilon) > 0$ and $c(\varepsilon) > 0$ such that, for any $x > x_0$ and for any $n \geq 1$,*

$$F^{*n}(x+\Delta) \leq c(\varepsilon)(1+\varepsilon)^n F(x+\Delta).$$

Proof. Let $\{\xi_n\}$ be a sequence of independent non-negative random variables with common distribution F. Put $S_n = \xi_1 + \ldots + \xi_n$. For $x_0 \geq 0$ and $k \geq 1$, put

$$A_n := A_n(x_0) = \sup_{x > x_0} \frac{F^{*n}(x+\Delta)}{F(x+\Delta)}.$$

Take any $\varepsilon > 0$. Appealing to Theorem 4.21, we conclude that x_0 may be chosen such that, for any $x > x_0$,

$$\mathbb{P}\{\xi_1 + \xi_2 \in x + \Delta, \xi_2 \leq x - x_0\} \leq (1 + \varepsilon/2)F(x + \Delta). \qquad (4.39)$$

For any $n > 1$ and $x > x_0$,

$$\mathbb{P}\{S_n \in x + \Delta\} = \mathbb{P}\{S_n \in x + \Delta, \xi_n \leq x - x_0\} + \mathbb{P}\{S_n \in x + \Delta, \xi_n > x - x_0\}$$
$$=: P_1(x) + P_2(x),$$

where, by the choice (4.39) of x_0 and by the definition of A_{n-1},

$$P_1(x) = \int_0^{x-x_0} \mathbb{P}\{S_{n-1} \in x - y + \Delta\}F(dy)$$

$$\leq A_{n-1} \int_0^{x-x_0} F(x - y + \Delta)F(dy)$$

$$= A_{n-1}\mathbb{P}\{\xi_1 + \xi_n \in x + \Delta, \xi_n \leq x - x_0\}$$

$$\leq A_{n-1}(1 + \varepsilon/2)F(x + \Delta). \qquad (4.40)$$

Further,

$$P_2(x) = \int_0^{x_0+T} \mathbb{P}\{\xi_n \in x - y + \Delta, \xi_n > x - x_0\}\mathbb{P}\{S_{n-1} \in dy\}$$
$$\leq \sup_{0 < t \leq x_0} F(x - t + \Delta).$$

Thus, if $x > 2x_0$, then

$$P_2(x) \leq L_1 F(x + \Delta),$$

where

$$L_1 = \sup_{0 < t \leq x_0, \, y > 2x_0} \frac{F(y - t + \Delta)}{F(y + \Delta)}.$$

If $x_0 < x \leq 2x_0$, then $P_2(x) \leq 1$ implies

$$\frac{P_2(x)}{F(x + \Delta)} \leq \frac{1}{\inf_{x_0 < x \leq 2x_0} F(x + \Delta)} =: L_2.$$

Since $F \in \mathcal{L}_\Delta$, both L_1 and L_2 are finite for x_0 sufficiently large. Put $L = \max(L_1, L_2)$. Then, for any $x > x_0$,

$$P_2(x) \leq LF(x + \Delta). \qquad (4.41)$$

It follows from (4.40) and (4.41) that $A_n \leq A_{n-1}(1 + \varepsilon/2) + L$ for $n > 1$. Therefore, the induction argument yields

$$A_n \leq A_1(1 + \varepsilon/2)^{n-1} + L \sum_{l=0}^{n-2}(1 + \varepsilon/2)^l \leq Ln(1 + \varepsilon/2)^{n-1}.$$

This implies the conclusion of the theorem. $\square$

4.7 Δ-Subexponential Distributions on the Real Line

As in the case of subexponential densities, for any fixed $\Delta = (0, T]$, there are two candidate conditions for the extension of the definition of Δ-subexponentiality to distributions on the whole real line. In this section we discuss these conditions briefly; the situation is very similar to that of Sect. 4.3, and we again refrain from making a judgement as to which condition is more appropriate. Thus our conditions, for a distribution F on $\mathbb{R}$, are:

(1) The distribution F^+ on $\mathbb{R}^+$ (given as usual by $F^+(x) = F(x)$ for $x \geq 0$ and $F^+(x) = 0$ for $x < 0$) is Δ-subexponential.

(2) The distribution F satisfies

$$F \in \mathcal{L}_\Delta \quad \text{and} \quad (F * F)(x + \Delta) \sim 2F(x + \Delta) \text{ as } x \to \infty. \tag{4.42}$$

Again the first of these conditions clearly preserves the tail property (since this holds for Δ-subexponentiality on $\mathbb{R}^+$), while the second does not. Further, analogously to the situation for subexponentiality of densities, these conditions are equivalent in the case of a distribution whose support is bounded below by some $a \in \mathbb{R}$.

We give here the analogues of Lemmas 4.12 and 4.13 for densities; the proofs are similarly analogous.

Lemma 4.26. *Suppose that the distribution F on $\mathbb{R}$ satisfies the condition* (4.42). *Then the distribution F^+ is Δ-subexponential.*

Lemma 4.27. *Suppose that the distribution F on $\mathbb{R}$ is such that, for some x_0 and $c < \infty$,*

$$F(x + y + \Delta) \leq cF(x + \Delta) \quad \text{for all } x > x_0 \text{ and } y > 0.$$

Suppose further that F^+ is Δ-subexponential. Then F satisfies the condition (4.42).

4.8 Sufficient Conditions for Δ-Subexponentiality

In this section, we give sufficient conditions for a distribution to be Δ-subexponentiality. There is much similarity between these conditions and those conditions given earlier for subexponentiality.

For a distribution F on $\mathbb{R}^+$ and $\Delta = (0, T]$, consider the new density $g(x)$ defined by

$$g(x) := c^{-1}F(x + \Delta), \qquad c = \int_0^\infty F(x + \Delta)\,dx.$$

If we prove that under certain conditions the density g is subexponential, then by (4.30) the distribution G with density g is Δ-subexponential. This implies, by Theorem 4.22, that F is also Δ-subexponential, since $F(x + \Delta) = cg(x) \sim cG(x + \Delta)/T$ as $x \to \infty$.

The latter observation gives us a simple way to prove the following two results.

Theorem 4.28. *Let the distribution F on $\mathbb{R}^+$ belong to the class $\mathcal{L}_\Delta$ where $\Delta = (0, T]$ for some finite $T > 0$. Suppose that there exist $c > 0$ and $x_0 < \infty$ such that $F(x + t + \Delta) \geq cF(x + \Delta)$ for any $t \in (0, x]$ and $x > x_0$. Then F is Δ-subexponential.*

Proof. The density g introduced above satisfies conditions of Theorem 4.14.   □

The Pareto distribution (with the tail $\overline{F}(x) = x^{-\alpha}$, $\alpha > 0$, $x \geq 1$) satisfies the conditions of Theorem 4.28. The same is true for any distribution F such that the function $F(x + \Delta)$ is regularly varying at infinity.

Theorem 4.29. *Suppose that the distribution F on $\mathbb{R}^+$ belongs to the class $\mathcal{L}_\Delta$ for some finite $\Delta = (0, T]$. Suppose also that for some x_0 the function $R(x) := -\ln F(x + \Delta)$ is concave for $x \geq x_0$. Suppose finally that there exists a function h such that $h(x) \to \infty$ as $x \to \infty$, $F(x + \Delta)$ is h-insensitive, and $xe^{-R(h(x))} \to 0$ as $x \to \infty$. Then F is Δ-subexponential.*

Proof. The density g introduced above satisfies conditions of Theorem 4.15.   □

To show the applicability of the latter theorem, we consider two examples. First, we consider the Weibull distribution F on the positive half-line $\mathbb{R}^+$ with tail function given by $\overline{F}(x) = e^{-x^\alpha}$, $x \geq 0$, $\alpha \in (0, 1)$, and let $\Delta = (0, T]$, for some finite T. Then it can be deduced that

$$F(x + \Delta) \sim \alpha T x^{\alpha - 1} \exp(-x^\alpha) \quad \text{as } x \to \infty.$$

From this, we want to show that $F(x + \Delta)$ is asymptotically equivalent to a function which satisfied the condition of Theorem 4.29. Thus, consider the distribution $\hat{F}$ with the tail function given by $\overline{\hat{F}}(x) = \min(1, x^{\alpha - 1}e^{-x^\alpha})$. Let x_0 be the unique positive solution to the equation $x^{1-\alpha} = e^{-x^\alpha}$. Then the function $\hat{g}(x) = -\ln \hat{F}(x + \Delta)$ is concave for $x \geq x_0$, and the conditions of Theorem 4.29 are satisfied with $h(x) = x^\gamma$, $\gamma \in (0, 1 - \alpha)$. Therefore, $\hat{F}$ is Δ-subexponential and, by Theorem 4.22, F is also Δ-subexponential.

Second, we consider the lognormal distribution F with the density f given by $f(x) = e^{-(\ln x - \mu)^2/2\sigma^2}/x\sqrt{2\pi\sigma^2}$ and note that the function

$$g(x) = -\ln(x^{-1}e^{-(\ln x - \mu)^2/2\sigma^2}) = \ln x + (\ln x - \mu)^2/2\sigma^2$$

is eventually concave. Since, for any fixed $\Delta = (0, T]$,

$$F(x + \Delta) \sim Tf(x)$$

as $x \to \infty$, the conditions of Theorem 4.29 are satisfied for any function h such that $h(x) = o(x)$. Thus F is Δ-subexponential.

We now show by two examples that the classes of Δ-subexponential distributions differ for different Δ and also the complexity of the relations between these classes. The first example deals with lattice distributions. Let the random variable ξ be positive and integer-valued, with $\mathbb{P}\{\xi = 2k\} = \gamma/k^2$ and $\mathbb{P}\{\xi = 2k+1\} = \gamma/2^k$, where γ is the appropriate normalizing constant. Then ξ has a lattice distribution F with span 1. By Theorem 4.28, F is $(0,2]$-subexponential. But it cannot be $(0,a]$-subexponential if a is not an even integer or infinity.

In the second example, we consider absolutely continuous distributions. Assume that ξ is the sum of two independent random variables: $\xi = \eta + \zeta$ where η is distributed uniformly on $(-1/8, 1/8)$ and $\mathbb{P}\{\zeta = k\} = \gamma/k^2$ for $k = 1, 2, \dots$ where γ is the appropriate normalising constant. Then the distribution F of ξ is absolutely continuous. It may be verified that F is $(0,1]$-subexponential, but cannot be $(0,a]$-subexponential if a is not an integer or infinity.

Finally, recall that in Sect. 4.6 we undertook to provide an example of a Δ-subexponential distribution F where the function $F(x + \Delta)$ is not asymptotically equivalent to any non-increasing function. Consider first a long-tailed function f such that $f(x) \in [1/x^2, 2/x^2]$ for all $x > 0$. Choose the function f in such a way that f is not asymptotically equivalent to a non-increasing function. For instance, one can define f as follows. Consider the increasing sequence $x_n = 2^{n/4}$. Put $f(x_{2n}) = 1/x_{2n}^2$ and $f(x_{2n+1}) = 2/x_{2n+1}^2$. Then assume that f is linear between any two consecutive members of the above sequence. Consider now the lattice distribution F on the set of natural numbers with $F\{n\} = f(n)$ for all sufficiently large integers n. Then by Theorem 4.22, F is Δ-subexponential, but $f(n) = F(n-1, n]$ is not asymptotically equivalent to a non-increasing function.

4.9 Local Asymptotics for a Randomly Stopped Sum

In this section, we give local analogues, both for subexponential densities and for Δ-subexponential distributions, of results which were given in Sect. 3.11 for subexponential distributions. We show that a random sum preserves a local subexponential property of independent identically distributed summands provided that the counting variable has a light-tailed distribution. We also establish the corresponding characteristic properties.

As is the case for the results obtained in Sect. 3.11, the results from this section are needed in a variety of models in which random sums may appear, including random walks, branching processes, and infinitely divisible laws.

We again consider a sequence $\xi, \xi_1, \xi_2, \dots$ of independent random variables with a common distribution F on $\mathbb{R}^+$ and their partial sums $S_0 = 0$, $S_n = \xi_1 + \dots + \xi_n$ for each $n \geq 1$, together a counting random variable τ which is independent of the sequence $\{\xi_n\}$ and takes values in $\mathbb{Z}^+$.

Density of a Randomly Stopped Sum

Let μ be either Lebesgue measure on $\mathbb{R}$ or counting measure on $\mathbb{Z}$. Throughout this section, the argument x of the density function f is either a real number if μ is Lebesgue measure; or an integer if μ is counting measure.

Theorem 4.30. *Let $\{p_n\}_{n\geq 1}$ be a non-negative sequence such that $\sum_{n\geq 1} p_n = 1$ and $m_p := \sum_{n\geq 1} n p_n$ is finite. Let the distribution F on $\mathbb{R}^+$ have a long-tailed density f with respect to μ. Define the density g on $\mathbb{R}^+$ by*

$$g(x) = \sum_{n\geq 1} p_n f^{*n}(x).$$

(i) *If the density f is subexponential and bounded, and if*

$$\sum_{n\geq 1} (1+\delta)^n p_n < \infty$$

for some $\delta > 0$, then

$$g(x) \sim m_p f(x) \quad as\ x \to \infty. \tag{4.43}$$

(ii) *If the relation (4.43) holds and $p_1 < 1$, then the density f is subexponential.*

Proof. The result (i) is immediate from Corollary 4.10, Theorem 4.11, and the dominated convergence theorem. We prove the second result. By Corollary 4.5, for any $k \geq 2$,

$$\liminf_{x\to\infty} f^{*k}(x)/f(x) \geq k.$$

If $p_1 < 1$ then $p_n > 0$ for some $n \geq 2$, and so, arguing as in the proof of Theorem 3.38, it follows from the above bound and from (4.43) that

$$\limsup_{x\to\infty} \frac{f^{*n}(x)}{f(x)} \leq n. \tag{4.44}$$

By Corollary 4.5, $f^{*(n-1)}$ is long-tailed and so, from (4.44) and Theorem 4.2,

$$n \geq \limsup_{x\to\infty} \frac{f^{*n}(x)}{f(x)} = \limsup_{x\to\infty} \frac{(f * f^{*(n-1)})(x)}{f(x)} \geq 1 + \limsup_{x\to\infty} \frac{f^{*(n-1)}(x)}{f(x)}.$$

It follows by induction from the above bound that

$$\limsup_{x\to\infty} \frac{f^{*2}(x)}{f(x)} \leq 2.$$

Again by Theorem 4.2, this implies that $\lim_{x\to\infty} f^{*2}(x)/f(x) = 2$, which implies the subexponentiality of the density f. $\qquad\square$

Δ-Subexponential Distributions and Random Sums

Analogously to Theorem 4.30, we have the following result.

Theorem 4.31. *Let $\Delta = (0,T]$ for some finite $T > 0$. Suppose that the distribution F on $\mathbb{R}^+$ is Δ-long-tailed ($F \in \mathcal{L}_\Delta$), and that the random variable τ (introduced at the start of Sect. 4.9) is such that $\mathbb{E}\tau < \infty$.*

(i) *If F is a Δ-subexponential distribution and if $\mathbb{E}(1+\delta)^\tau < \infty$ for some $\delta > 0$, then*

$$\frac{\mathbb{P}\{S_\tau \in x + \Delta\}}{F(x+\Delta)} \to \mathbb{E}\tau \quad as \; x \to \infty. \tag{4.45}$$

(ii) *If $\mathbb{P}\{\tau > 1\} > 0$ and further the relation (4.45) holds, then the distribution F is Δ-subexponential.*

Proof. The proof of (i) follows from Corollary 4.24, Theorem 4.25, and the dominated convergence theorem. We prove (ii). Since $F \in \mathcal{L}_\Delta$, it follows from Corollary 4.19 that, for any $k \geq 2$,

$$\liminf_{x\to\infty} \frac{F^{*k}(x+\Delta)}{F(x+\Delta)} \geq k. \tag{4.46}$$

If $\mathbb{P}\{\tau = n\} > 0$ for some $n \geq 2$, then, again arguing as in the proof of Theorem 3.38, it follows from the above bound and from (4.45) that

$$\limsup_{x\to\infty} \frac{F^{*n}(x+\Delta)}{F(x+\Delta)} \leq n. \tag{4.47}$$

Since $F \in \mathcal{L}_\Delta$, by Corollary 4.19 the convolution $F^{*(n-1)}$ also belongs to the class $\mathcal{L}_\Delta$. Hence, by (4.47) and Theorem 4.17,

$$n \geq \limsup_{x\to\infty} \frac{F^{*n}(x+\Delta)}{F(x+\Delta)}$$

$$= \limsup_{x\to\infty} \frac{(F * F^{*(n-1)})(x+\Delta)}{G(x+\Delta)}$$

$$\geq 1 + \limsup_{x\to\infty} \frac{F^{*(n-1)}(x+\Delta)}{F(x+\Delta)}.$$

It follows by induction from the above bound that

$$\limsup_{x\to\infty} \frac{F^{*2}(x+\Delta)}{F(x+\Delta)} \leq 2.$$

Again by Theorem 4.17, this implies that $\lim_{x\to\infty} F^{*2}(x+\Delta)/F(x+\Delta) = 2$, which implies the Δ-subexponentiality of the distribution F. $\qquad\square$

4.10 Local Subexponentiality of Integrated Tails

In the present section we answer the question of what conditions are necessary for the integrated tail distribution to have a subexponential density and to be Δ-subexponential.

Theorem 4.32. *Let the distribution F on $\mathbb{R}^+$ be long-tailed. Then the following statements are equivalent:*
 (i) *F belongs to the class $\mathcal{S}^*$.*
 (ii) *The density of the integrated tail distribution F_I is subexponential.*
 (iii) *F_I is Δ-subexponential, for any $\Delta = (0,T]$, $T > 0$.*
 (iv) *F_I is Δ-subexponential, for some $\Delta = (0,T]$, $T > 0$.*

Proof. The distribution F_I is bounded from below. On the support of F_I, its density equals $\overline{F}(x)$. Since F is long-tailed, the density of F_I is long-tailed as well. Then, by Theorem 4.7, the density of F_I is subexponential if and only if, for every function h such that $h(x) \to \infty$ as $x \to \infty$,

$$\int_{h(x)}^{x-h(x)} \overline{F}(x-y)\overline{F}(y)dy = o(\overline{F}(x)) \text{ as } x \to \infty,$$

which is equivalent to $F \in \mathcal{S}^*$, see Theorem 3.24. Hence (i) is equivalent to (ii).

Since F is long-tailed, $F_I(x,x+\Delta] \sim T\overline{F}(x)$ as $x \to \infty$. In particular, the function $F_I(x,x+\Delta]$ is long-tailed and

$$\int_{h(x)}^{x-h(x)} F_I(x+\Delta - y)F_I(dy) = \int_{h(x)}^{x-h(x)} F_I(x+\Delta - y)\overline{F}(y)dy$$

$$\sim T\int_{h(x)}^{x-h(x)} F(x-y)\overline{F}(y)dy \quad \text{as } x \to \infty.$$

This shows, via Theorem 4.21, why both (iii) and (iv) are equivalent to (i). $\Box$

4.11 Comments

Local theorems for some classes of lattice distributions are given by Chover, Ney, and Wainger in [14, Sect. 2]. Densities are considered in [14, Sect. 2] (requiring continuity) and by Klüppelberg in [33] who considered asymptotics of densities for a special case (see also Sgibnev [49] for some results on densities on $\mathbb{R}$).

Much of the material of this chapter is adapted from the paper by Asmussen, Foss, and Korshunov [4].

4.12 Problems

4.1. Prove by definition that the Cauchy density is subexponential in the sense of (4.18).

4.2. Prove by direct estimations for the convolution that the Pareto density is subexponential.

4.3. Prove by direct estimations for convolution that any regularly varying at infinity density is subexponential.

4.4. Let f and g be two regularly varying at infinity densities and $0 < p < 1$. Prove that the density $pf + (1 - p)g$ is subexponential too.

4.5. Let F have a subexponential density with respect to the counting measure on $\mathbb{Z}^+$ and G be the uniform distribution in the interval $[a, b]$. Prove that the density of the convolution $F * G$ is subexponential if and only if $b - a$ is an integer.

4.6. Let F have a subexponential density with respect to the counting measure on $\mathbb{Z}^+$ and G be the exponential distribution. Prove that the density of the convolution $F * G$ is not subexponential. Is this convolution $[0, 1)$-subexponential?

4.7. Let F be a distribution on $\mathbb{R}^+$ with subexponential density with respect to the Lebesgue measure. Let G have either

(i) a distribution with a compact support or
(ii) the Poisson distribution.

Prove that the density of the convolution $F * G$ is subexponential.

4.8. Let F be a $[0, 1)$-subexponential distribution on $\mathbb{R}^+$. Let G be the uniform distribution in the interval $[a, b]$ where $b - a$ is an integer. Prove that the density of the convolution $F * G$ is subexponential.

4.9. Suppose that f and g are weakly tail-equivalent long-tailed densities on $\mathbb{R}^+$. Prove that the convolutions $f * f$ and $g * g$ are weakly tail-equivalent too. Prove a similar result for the n-fold convolutions, $n \geq 3$.

4.10. Suppose that $\xi_1, \ldots, \xi_n$ are independent random variables with common distribution density f. Prove that the distribution density of the maximum, $\max(\xi_1, \ldots, \xi_n)$, is subexponential if and only if f is subexponential.

4.11. Suppose that $\xi_1, \ldots, \xi_n$ are independent random variables with a common exponential distribution. Find the asymptotics, as $x \to \infty$, for the probability density of the sum $\xi_1^{\alpha_1} + \ldots + \xi_n^{\alpha_n}$ if

(i) all $\alpha_i > 1$;
(ii) all $\alpha_i < 0$;
(iii) all $\alpha_i \notin [0, 1]$.

4.12. Let independent random variables $\xi_1, \ldots, \xi_n$ have the standard Cauchy distribution, and let a counting random variable τ to be independent of the ξ's. Compute the density of S_τ.

4.13. In the conditions of the previous problem, prove that the relation

$$\mathbb{P}\{S_\tau > x\} \sim \mathbb{E}\tau \mathbb{P}\{\xi_1 > x\} \quad \text{as } x \to \infty$$

holds for every τ with a finite mean.

4.14. Let $X(t)$ be a compound Poisson process such that its jumps have a subexponential distribution density f. For every t, find the asymptotic behaviour of the distribution density of $X(t)$ in terms of f.

4.15. Prove subexponentiality of the distribution density of the product $\xi_1 \xi_2$ of two independent random variables with common exponential distribution.

4.16. Prove that the distribution density of the product $\xi_1 \xi_2$ of two independent random variables with common normal distribution is not subexponential. Prove subexponentiality of the distribution density of the product of three independent normal variables.

4.17. Prove that the distribution of the product $\xi_1 \xi_2$ of two independent random variables with common Poisson distribution is heavy-tailed. Prove that its density with respect to the counting measure is not long-tailed and, therefore, is not subexponential.

4.18. Prove that, for any two independent non-negative light-tailed random variables on $\mathbb{Z}^+$, the distribution density (with respect to the counting measure) of their product cannot be long-tailed.

4.19. In the conditions of Problem 2.24, assume that the limit $\lim_{i \to \infty} p_{ii}$ exists and is less than 1, and that $\limsup_{i \to \infty} (p_{i,i+1} - p_{i,i-1}) = 0$. Show that then the invariant distribution is locally long-tailed. What kind of a regular behaviour of the transition probabilities has to be assumed to ensure that the invariant probabilities π_i are

(i) regularly varying at infinity (as $i \to \infty$);
(ii) varying at infinity in Weibullian way.

4.20. *Excess process.* In the conditions of Problem 2.26, assume that F is long-tailed. Prove that then the invariant distribution is locally subexponential if and only if the distribution F is strong subexponential.

4.21. In the conditions of Problem 2.27, what extra conditions should be assumed for the invariant distribution to be locally subexponential?

4.22. *Return time.* Suppose S_n is a random walk in $\mathbb{Z}^d$, $d \geq 3$, with zero drift and with a finite covariance matrix B. It is known that this random walk is transient with $p := \mathbb{P}\{\tau_1 < \infty\} < 1$ where $\tau_1 := \min\{n \geq 1 : S_n = 0\}$ is the time of the first return to the origin. It is also known that $\mathbb{P}\{\tau_1 = n\} \sim cn^{-d/2}$ as $n \to \infty$. Find the coefficient c.

Chapter 5
Maximum of Random Walk

In this chapter, we study a random walk whose increments have a (right) heavy-tailed distribution with a negative mean. We also consider applications to queueing and risk processes.

The maximum of such a random walk is almost surely finite, and our interest is in the tail asymptotics of the distribution of this maximum, for both infinite and finite time horizons; we are further interested in the local asymptotics for the maximum in the case of an infinite time horizon. We use direct probabilistic techniques and show that, under the appropriate subexponentiality conditions, the main reason for the maximum to be far away from zero is again that a single increment of the walk is similarly large.

We present here two approaches for deriving such results, the first using a first renewal time at which the random walk exceeds a "tilted" level and the second using classical ladder epochs and heights. It turns out that the former approach is more direct since it is based on more elementary arguments, and we start with it in Sects. 5.1 and 5.2. In Sect. 5.1 we deal with the infinite time horizon and first obtain a general lower bound, and then the correct asymptotics, for the distribution of the maximum. Similar results for finite time horizons (with uniformity in time) are given in Sect. 5.2.

We then turn to the classical ladder heights approach. This allows us to obtain both tail and local asymptotics for the maximum of the random walk in the case of an infinite time horizon. In Sects. 5.3 and 5.4 we recall known basic results on the ladder structure and on taboo renewal measures. In Sect. 5.5 we give results on bounds and asymptotics for both the tail and the local probabilities of the first ascending ladder height; this will lead to another proof of the tail asymptotics for the infinite-time maximum of the walk (Sect. 5.6) and also to the asymptotics for the local probabilities of the distribution of the maximum (Sect. 5.7). In Sect. 5.8 we present the asymptotics for the density of the maximum in the case where the density exists; this is also based on the ladder-heights representation. In each of three Sects. 5.6–5.8, we show that the corresponding condition on the distribution to belong to the appropriate class is not only sufficient but also necessary, for the desired asymptotics to hold. In Sect. 5.9, we consider the three particular cases where the distribution of the first strictly ascending ladder height may be explicitly calculated and provide new local theorems and improved bounds.

The next Sects. 5.10–5.12 are devoted to applications. In Sect. 5.10 we obtain the asymptotics for the stationary waiting time distribution in a stable single-server queue with subexponential type distribution of service times. In Sect. 5.11 we consider the classical Cramér–Lundberg model of the collective theory of risk and, in particular, find the asymptotics for the ruin probability, both in infinite and finite time horizons. In Sect. 5.12 we demonstrate how subexponential distributions appear in the theory of branching processes.

Finally, in Sect. 5.13, for standard cases, we formulate and prove a limit theorem for the distribution of the quadruple that includes the time to exceed a high level by a random walk, the position at this time, the position at the prior time and the trajectory up to it.

5.1 Asymptotics for the Maximum of a Random Walk with a Negative Drift

We give an elementary probabilistic description of the asymptotic behaviour of the distribution of the maximum of a random walk with negative drift and heavy-tailed increments (see Theorem 5.2 below). The underlying intuition of the result is that the only significant way in which a large value of the maximum can be attained is through "one big jump" by the random walk away from its mean path. We give here a relatively short proof from first principles which captures this intuition. It is similar in spirit to the probabilistic proof related to the ladder heights (which may also be of use for deriving local asymptotics), but by considering instead a first renewal time at which the random walk exceeds a "tilted" level, the argument becomes more elementary. In particular, subsequent renewals have an asymptotically negligible probability under appropriate limits, and results from renewal theory—notably the derivation and use of the Pollaczeck–Khinchine formula—are not required.

We proceed with the proof by deriving separately the lower and the upper bounds, since no restrictions (apart of the negativeness of the mean!) are required for the former to hold while subexponentiality is needed for the latter.

Let $\xi_1, \xi_2, \ldots$ be independent identically distributed random variables with distribution function F such that $\mathbb{E}\xi_1 = -a < 0$. Let $S_0 = 0$, $S_n = \xi_1 + \ldots + \xi_n$ for $n \geq 1$ be the associated random walk. Let $M_n = \max(S_i, 0 \leq i \leq n)$ for $n \geq 0$ be the finite time horizon maximum of the random walk S_n and let $M = \sup(S_n, n \geq 0)$ be its global maximum. It follows from the strong law of large numbers that $\mathbb{P}\{M < \infty\} = 1$, and our interest in this section is in the distribution of M.

We start with the lower bound, which is proved by a quite elementary equilibrium identity.

Theorem 5.1. *Suppose that* $\mathbb{E}\xi_1 = -a < 0$. *Then, for any* $x \geq 0$,

$$\mathbb{P}\{M > x\} \geq \frac{\int_x^\infty \overline{F}(y)dy}{a + \int_x^\infty \overline{F}(y)dy},$$

and, in particular,

$$\liminf_{x \to \infty} \frac{\mathbb{P}\{M > x\}}{\overline{F}_I(x)} \geq \frac{1}{a}.$$

Proof. Let ξ be a random variable with distribution F which is independent of M. Then M has the same distribution as $(M + \xi)^+ := \max(0, M + \xi)$. Now fix $x \geq 0$. For $z > 0$ consider the function

$$L_z(y) = \begin{cases} x & \text{if } y \leq x, \\ y & \text{if } y \in (x, x+z], \\ x+z & \text{if } y > x+z. \end{cases}$$

Since this function is bounded, $\mathbb{E}L_z(M)$ is finite and $\mathbb{E}L_z(M) = \mathbb{E}L_z(M + \xi)$. Therefore,

$$\mathbb{E}(L_z(M + \xi) - L_z(M)) = 0.$$

We have $|L_z(M + \xi) - L_z(M)| \leq |\xi|$ for all z and $L_z(M + \xi) - L_z(M) \to L(M + \xi) - L(M)$ as $z \to \infty$ where

$$L(y) = \begin{cases} x & \text{if } y \leq x, \\ y & \text{if } y > x. \end{cases}$$

Hence, by dominated convergence, we obtain the equality

$$\mathbb{E}(L(M + \xi) - L(M)) = 0. \tag{5.1}$$

We make use of the following bounds. For $y \in [0, x]$,

$$L(y + \xi) - L(y) = (y + \xi - x)\mathbb{I}\{y + \xi > x\} \geq (\xi - x)\mathbb{I}\{\xi > x\},$$

and so

$$\mathbb{E}\{L(M + \xi) - L(M); M \leq x\} \geq \mathbb{E}\{\xi - x; \xi > x\}\mathbb{P}\{M \leq x\}. \tag{5.2}$$

For $y > x$,

$$L(y + \xi) - L(y) \geq \xi,$$

and so

$$\mathbb{E}\{L(M + \xi) - L(M); M > x\} \geq \mathbb{E}\xi\mathbb{P}\{M > x\}. \tag{5.3}$$

Substituting (5.2) and (5.3) into (5.1) we get the inequality

$$\mathbb{E}\{\xi - x; \xi > x\}\mathbb{P}\{M \leq x\} \leq -\mathbb{E}\xi\mathbb{P}\{M > x\}.$$

Therefore,

$$\mathbb{P}\{M > x\} \geq \frac{\mathbb{E}\{\xi - x; \xi > x\}}{a + \mathbb{E}\{\xi - x; \xi > x\}} = \frac{\int_x^\infty \overline{F}(y)dy}{a + \int_x^\infty \overline{F}(y)dy},$$

where the final equality follows from (2.23).   □

We now give our main result of this section, for the asymptotic behaviour of the tail of M.

Theorem 5.2. *Suppose that, in addition to the condition* $\mathbb{E}\xi_1 = -a < 0$*, the integrated tail distribution* F_I *is subexponential. Then*

$$\mathbb{P}\{M > x\} \sim a^{-1}\overline{F}_I(x) \quad \text{as } x \to \infty.$$

Proof. By Theorem 5.1, it is sufficient to establish the upper bound associated with the required asymptotics. Given $\varepsilon > 0$ and some (eventually large) $A > a$, define renewal times $0 =: \tau_0 < \tau_1 \leq \tau_2 \leq \dots$ for the process $\{S_n\}$ by

$$\tau_1 = \min\{j \geq 1 : S_j > A - j(a - \varepsilon)\} \leq \infty$$

(here we make the standard convention $\min \varnothing = \infty$), and, for $k \geq 2$,

$$\tau_k = \infty, \quad \text{if } \tau_{k-1} = \infty,$$
$$\tau_k = \tau_{k-1} + \min\{j \geq 1 : S_{\tau_{k-1}+j} - S_{\tau_{k-1}} > A - j(a - \varepsilon)\}, \quad \text{if } \tau_{k-1} < \infty.$$

Observe that, for any k, the joint distribution of the vectors

$$(\tau_1, S_{\tau_1}), (\tau_2 - \tau_1, S_{\tau_2} - S_{\tau_1}), \dots, (\tau_k - \tau_{k-1}, S_{\tau_k} - S_{\tau_{k-1}}), \tag{5.4}$$

conditioned on $\tau_k < \infty$, is that of independent identically distributed vectors. Since $\mathbb{E}\xi_1 < 0$, by the strong law of large numbers,

$$\gamma := \mathbb{P}\{\tau_1 < \infty\} \to 0 \quad \text{as } A \to \infty. \tag{5.5}$$

Define also $S_\infty = -\infty$. Since $\tau_1 = n$ implies $S_{n-1} \leq A - (n-1)(a - \varepsilon)$, we now have that, for all sufficiently large x,

$$\mathbb{P}\{S_{\tau_1} > x\} = \sum_{n=1}^\infty \mathbb{P}\{\tau_1 = n, S_n > x\}$$

$$\leq \sum_{n=1}^\infty \mathbb{P}\{S_{n-1} \leq A - (n-1)(a - \varepsilon), S_n > x\}$$

$$\leq \sum_{n=1}^\infty \mathbb{P}\{\xi_n > x - A + (n-1)(a - \varepsilon)\}.$$

Therefore, again for all sufficiently large x,

$$\mathbb{P}\{S_{\tau_1} > x\} \le \sum_{n=0}^{\infty} \overline{F}(x - A + n(a - \varepsilon)) \le \frac{1}{a - \varepsilon} \overline{F}_I(x - A - a + \varepsilon), \quad (5.6)$$

where the second inequality above follows from this observation: for any n,

$$\overline{F}(x - A + n(a - \varepsilon)) \le \frac{1}{a - \varepsilon} \int_{x - A + (n-1)(a-\varepsilon)}^{x - A + n(a-\varepsilon)} \overline{F}(y) dy.$$

Let φ_1, φ_2, ... be independent identically distributed random variables having tail distribution

$$\mathbb{P}\{\varphi_1 > x\} = \mathbb{P}\{S_{\tau_1} > x \mid \tau_1 < \infty\}, \quad x \in \mathbb{R}.$$

Then, from (5.6) and since F_I is long-tailed,

$$\mathbb{P}\{\varphi_1 > x\} \le \overline{G}(x), \quad x \in \mathbb{R}, \quad (5.7)$$

for some distribution function G on $\mathbb{R}$ satisfying

$$\lim_{x \to \infty} \frac{\overline{G}(x)}{\overline{F}_I(x)} = \frac{1}{\gamma(a - \varepsilon)}. \quad (5.8)$$

It follows from the subexponentiality of F_I and Corollary 3.13 that the distribution G is subexponential. Thus, by applying Theorem 3.37 with a geometrically distributed independent stopping time, we have

$$(1 - \gamma) \sum_{k=0}^{\infty} \gamma^k \overline{G^{*k}}(x) \sim \frac{\gamma}{1 - \gamma} \overline{G}(x) \quad \text{as } x \to \infty.$$

From the stochastic majorisation (5.7) and the relation (5.8), we now get the following asymptotic upper bound:

$$\sum_{k=1}^{\infty} \gamma^k \mathbb{P}\{\varphi_1 + \ldots + \varphi_k > x\} \le \frac{\gamma + o(1)}{(1 - \gamma)^2} \overline{G}(x)$$

$$\le \frac{1 + o(1)}{(1 - \gamma)^2 (a - \varepsilon)} \overline{F}_I(x) \quad \text{as } x \to \infty. \quad (5.9)$$

If $M > x$ then there exist τ_k and $j \in [\tau_k, \tau_{k+1})$ such that $S_j > x$. Then necessarily $S_{\tau_k} > x - A + a - \varepsilon$. (To see this assume that, on the contrary, $S_{\tau_k} \le x - A + a - \varepsilon < x$. Then $\tau_k < j < \tau_{k+1}$ and $S_j - S_{\tau_k} > x - (x - A + a - \varepsilon) = A - a + \varepsilon$. Hence we have the contradiction that $\tau_{k+1} \le j$.) It follows that

$$\{M > x\} \subseteq \bigcup_{k=1}^{\infty} \{S_{\tau_k} > x - A + a - \varepsilon\}.$$

We thus have (again for sufficiently large x) that

$$\mathbb{P}\{M > x\} \leq \sum_{k=1}^{\infty} \mathbb{P}\{S_{\tau_k} > x - A + a - \varepsilon\}$$

$$\leq \sum_{k=1}^{\infty} \gamma^k \mathbb{P}\{\varphi_1 + \ldots + \varphi_k > x - A + a - \varepsilon\},$$

by (5.4) and by the construction of the random variables φ_i. Also using (5.9), we now have

$$\limsup_{x \to \infty} \frac{\mathbb{P}\{M > x\}}{\overline{F}_I(x)} \leq \frac{1}{(a - \varepsilon)(1 - \gamma)^2}.$$

Now let $A \to \infty$, so that $\gamma \to 0$ by (5.5), and then let $\varepsilon \to 0$ to obtain the required upper bound:

$$\limsup_{x \to \infty} \frac{\mathbb{P}\{M > x\}}{\overline{F}_I(x)} \leq \frac{1}{a}. \qquad \square$$

It seems to be tempting to complement the equivalence relation from Theorem 5.2 and the lower bound from Theorem 5.1 by an upper bound. There have been many attempts to find such bounds, either for the tail distribution of the maximum of a random walk or for the tail distribution of a geometric sum of independent identically distributed positive random variables, see, e.g. [30]. However, there are no satisfactory solutions or approaches, and the problem is still open, see, e.g. [38].

5.2 Finite Time Horizon Asymptotics

We continue to study the random walk with negative drift introduced in the previous section. Recall that $M_n = \max(S_i, 0 \leq i \leq n)$ is defined to be the maximum of the random walk to time n. In this section we derive asymptotics, uniform in n, for the probability $\mathbb{P}\{M_n > x\}$ as $x \to \infty$ under heavy-tailedness assumptions. If F is (whole-line) subexponential and n is fixed then, since subexponentiality is a tail property and by Theorem 2.40, the inequalities

$$S_n \leq M_n \leq \sum_{k=1}^{n} \xi_k^+$$

imply that
$$\mathbb{P}\{M_n > x\} \sim n\overline{F}(x) \qquad \text{as } x \to \infty.$$

In the next theorem we produce asymptotics which are uniform in n. The underlying intuition of the result is again that the only significant way in which a high value of the partial maximum can be attained is via a "big jump" of one of its increments. The proof of the lower bound is based on direct computations and requires the extra assumption of long-tailedness of the distribution F of the increments ξ_i. The proof of the upper bound is similar to that of Theorem 5.2, although the condition of that theorem that F_I be subexponential requires to be strengthened slightly to $F \in \mathcal{S}^*$ (see Sect. 3.4 and in particular Theorem 3.27).

In what follows we write $f(x,n) \geq (1 + o(1))g(x,n)$ as $x \to \infty$ uniformly in $n \geq 1$ if

$$\liminf_{x\to\infty} \inf_{n\geq 1} \frac{f(x,n)}{g(x,n)} \geq 1,$$

and we write $f(x,n) \sim g(x,n)$ as $x \to \infty$ uniformly in $n \geq 1$ if

$$\sup_{n\geq 1}\left|\frac{f(x,n)}{g(x,n)} - 1\right| \to 0 \text{ as } x \to \infty.$$

Theorem 5.3. *Let $\mathbb{E}\xi = -a < 0$. Suppose that the distribution F is long-tailed ($F \in \mathcal{L}$). Then*

$$\mathbb{P}\{M_n > x\} \geq \frac{1 + o(1)}{a} \int_x^{x+na} \overline{F}(y)dy \quad \text{as } x \to \infty, \text{ uniformly in } n \geq 1. \quad (5.10)$$

If, in addition, the distribution is strong subexponential ($F \in \mathcal{S}^*$), then

$$\mathbb{P}\{M_n > x\} \sim \frac{1}{a} \int_x^{x+na} \overline{F}(y)dy \quad \text{as } x \to \infty, \text{ uniformly in } n \geq 1. \quad (5.11)$$

Proof. We prove first the lower bound given in (5.10). Since $\mathbb{E}\xi_1 < 0$, it follows from the strong law of large numbers that, given $\varepsilon > 0$ and $\delta > 0$, we can choose A sufficiently large that

$$\mathbb{P}\{|S_k + ka| \leq A + k\varepsilon \text{ for all } k \leq n\} \geq 1 - \delta \quad \text{for all } n \geq 0. \quad (5.12)$$

Then the following lower bound is immediate:

$$\mathbb{P}\{M_n > x\} = \sum_{k=0}^{n-1} \mathbb{P}\{M_k \leq x, S_{k+1} > x\}$$

$$\geq \sum_{k=0}^{n-1} \mathbb{P}\{M_k \leq x, |S_j + ja| \leq A + j\varepsilon \text{ for all } j \leq k,$$

$$\xi_{k+1} > x + A + k(a + \varepsilon)\}.$$

By the independence of random variables ξ_i and by (5.12), we have

$$\mathbb{P}\{M_n > x\} \geq \sum_{k=0}^{n-1} \mathbb{P}\{M_k \leq x, |S_j + ja| \leq A + j\varepsilon \text{ for all } j \leq k\}$$

$$\times \mathbb{P}\{\xi_{k+1} > x + A + k(a + \varepsilon)\}$$

$$\geq \sum_{k=0}^{n-1} (1 - 2\delta)\overline{F}(x + A + k(a + \varepsilon)),$$

where the last inequality holds for all x sufficiently large that

$$\mathbb{P}\{M > x\} \leq \delta \tag{5.13}$$

which implies that $\mathbb{P}\{M_k > x\} \leq \delta$ for all k. By summation of the inequalities

$$\overline{F}(x + A + k(a + \varepsilon)) \geq \frac{1}{a + \varepsilon} \int_{x + k(a + \varepsilon)}^{x + (k+1)(a + \varepsilon)} \overline{F}(y + A) dy,$$

we get

$$\mathbb{P}\{M_n > x\} \geq \frac{1 - 2\delta}{a + \varepsilon} \int_x^{x + n(a + \varepsilon)} \overline{F}(y + A) dy.$$

Since F is assumed to be long-tailed, it now follows that

$$\mathbb{P}\{M_n > x\} \geq \frac{1 - 3\delta}{a + \varepsilon} \int_x^{x + n(a + \varepsilon)} \overline{F}(y) dy$$

for all x sufficiently large that (5.13) holds. That the inequality (5.10) holds with the required uniformity in n now follows by letting $\delta, \varepsilon \to 0$.

We now prove (5.11). Here F is assumed to belong to the class $\mathcal{S}^*$, so it is in particular long-tailed. Hence, it is sufficient to establish the upper bound in (5.11). Given $\varepsilon > 0$ and $A > a$, define renewal times $0 =: \tau_0 < \tau_1 \leq \tau_2 \leq \ldots$ for the process $\{S_k\}$ as in the proof of Theorem 5.2.

Analogously to (5.6), we obtain

$$\mathbb{P}\{S_{\tau_1 \wedge n} > x\} \leq \sum_{k=0}^{n-1} \overline{F}(x - A + k(a - \varepsilon)) \leq \frac{1}{a - \varepsilon} \int_x^{x + na} \overline{F}(y - A - a + \varepsilon) dy.$$

Since F is long-tailed,

$$\mathbb{P}\{S_{\tau_1 \wedge n} > x\} \leq \frac{1 + \varepsilon}{a - \varepsilon} \int_x^{x + na} \overline{F}(y) dy \tag{5.14}$$

for all sufficiently large x uniformly in $n \geq 1$. This means that we can choose x_0 such that (5.14) holds for all $x \geq x_0$ and for all $n = 1, 2, \ldots$.

Let $\varphi_{n,1}, \varphi_{n,2}, \ldots$ be independent identically distributed random variables such that

$$\mathbb{P}\{\varphi_{n,1} > x\} = \mathbb{P}\{S_{\tau_1 \wedge n} > x \mid \tau_1 < \infty\}, \quad x \in \mathbb{R}.$$

Then, from (5.14), for $x \geq x_0$,

$$\mathbb{P}\{\varphi_{n,1} > x\} \leq \int_x^{x + na} \overline{G}_n(y) dy, \quad x \in \mathbb{R}, n \geq 1, \tag{5.15}$$

for some distribution function G_n on $\mathbb{R}$ satisfying

$$\lim_{x \to \infty} \frac{\overline{G}_n(x)}{\overline{F}(x)} = \frac{1+\varepsilon}{\gamma(a-\varepsilon)}. \tag{5.16}$$

From the condition $F \in \mathcal{S}^*$ and Corollary 3.26, we have $G_n \in \mathcal{S}^*$. We may now apply Corollary 3.40 with a geometrically distributed stopping time to obtain that

$$(1-\gamma) \sum_{k=0}^{\infty} \gamma^k \overline{G_n^{*k}}(x) \sim \frac{\gamma}{1-\gamma} \overline{G}_n(x)$$

as $x \to \infty$ uniformly in $n \geq 1$. Using also the conditions (5.15) and (5.16), we get the following asymptotic upper bound:

$$\sum_{k=1}^{\infty} \gamma^k \mathbb{P}\{\varphi_{n,1} + \ldots + \varphi_{n,k} > x\} \leq \frac{\gamma + o(1)}{(1-\gamma)^2} \overline{G}_n(x)$$

$$\leq \frac{1+\varepsilon+o(1)}{(1-\gamma)^2(a-\varepsilon)} \int_x^{x+na} \overline{F}(y)dy \tag{5.17}$$

as $x \to \infty$ uniformly in $n \geq 1$.

If $M_n > x$, then there exist $\tau_k \leq n$ and $j \in [\tau_k, \tau_{k+1})$ such that $S_j > x$. Then, exactly as in the proof of Theorem 5.2, we have that necessarily $S_{\tau_k} > x - A + a - \varepsilon$. It follows that

$$\{M_n > x\} \subseteq \bigcup_{k=1}^{\infty} \{S_{\tau_k \wedge n} > x - A + a - \varepsilon\}.$$

Therefore,

$$\mathbb{P}\{M_n > x\} \leq \sum_{k=1}^{\infty} \mathbb{P}\{S_{\tau_k \wedge n} > x - A + a - \varepsilon\}$$

$$\leq \sum_{k=1}^{\infty} \gamma^k \mathbb{P}\{\varphi_{n,1} + \ldots + \varphi_{n,k} > x - A + a - \varepsilon\},$$

by the construction of the random variables φ_i. Using (5.17) we obtain

$$\limsup_{x \to \infty} \sup_{n \geq 1} \frac{\mathbb{P}\{M_n > x\}}{\int_x^{x+na} \overline{F}(y)dy} \leq \frac{1+\varepsilon}{(a-\varepsilon)(1-\gamma)^2}.$$

Now first let $A \to \infty$, so that $\gamma \to 0$ by (5.5). Then let $\varepsilon \to 0$ to obtain the required upper bound

$$\limsup_{x \to \infty} \sup_{n \geq 1} \frac{\mathbb{P}\{M_n > x\}}{\int_x^{x+na} \overline{F}(y)dy} \leq \frac{1}{a},$$

which, together with the lower bound (5.10), implies the required uniform asymptotics (5.11). $\qquad\square$

We conclude this section with two theorems that are nothing other than versions of the principle of a single big jump for M and M_n. For any $A > 0$ and $\varepsilon > 0$ consider events

$$B_k := \left\{ |S_j + aj| \le j\varepsilon + A \text{ for all } j \le k, \ \xi_{k+1} > x + ka \right\}$$

which, for large x, roughly speaking means that up to time k the random walk S_j moves down according to the strong law of large numbers and then makes a big jump up. As stated in the next theorem, the union of these events describes the most probable way by which large deviations of M or M_n can occur.

Theorem 5.4. *Let* $\mathbb{E}\xi = -a < 0$ *and* $F_I \in \mathcal{S}$. *Then, for any fixed* $\varepsilon > 0$,

$$\lim_{A \to \infty} \lim_{x \to \infty} \mathbb{P}\{\cup_{k=0}^{\infty} B_k | M > x\} = 1.$$

If, in addition, $F \in \mathcal{S}^*$, *then, for any fixed* $\varepsilon > 0$,

$$\lim_{A \to \infty} \lim_{x \to \infty} \inf_{n \ge 1} \mathbb{P}\{\cup_{k=0}^{n-1} B_k | M_n > x\} = 1.$$

Proof. We prove the second assertion only, since the proof of the first is similar.

Since, for $k \le n - 1$, each of the events

$$\tilde{B}_k := \left\{ |S_j + aj| \le j\varepsilon + A \text{ for all } j \le k, \ M_k \le x, \ \xi_{k+1} > x + A + k(a + \varepsilon) \right\}$$

is contained in B_k and implies that $S_k > x$ so that $M_n > x$, we consequently have that

$$\mathbb{P}\{\cup_{k=0}^{n} B_k | M_n > x\} \ge \mathbb{P}\{\cup_{k=0}^{n-1} \tilde{B}_k | M_n > x\} = \frac{\mathbb{P}\{\cup_{k=0}^{n-1} \tilde{B}_k\}}{\mathbb{P}\{M_n > x\}}. \tag{5.18}$$

The events $\tilde{B}_k$ are disjoint, hence

$$\mathbb{P}\{\cup_{k=0}^{n-1} \tilde{B}_k\} = \sum_{k=0}^{n-1} \mathbb{P}\{\tilde{B}_k\}.$$

As was shown in the proof of the previous theorem, for any fixed $\delta > 0$, there exists A such that, for all $x > A$,

$$\mathbb{P}\{\cup_{k=1}^{n-1} \tilde{B}_k\} \ge \frac{1-\delta}{a+\varepsilon} \int_x^{x+n(a+\varepsilon)} \overline{F}(y)\,dy.$$

Substituting this estimate and the asymptotics for M_n into (5.18) we deduce that

$$\lim_{x \to \infty} \inf_{n \ge 1} \mathbb{P}\{\cup_{k=0}^{n} B_k | M_n > x\} \ge \frac{(1-\delta)a}{a+\varepsilon}.$$

Now we can make $\delta > 0$ as small as we please by choosing a sufficiently large A. Therefore,

$$\lim_{A \to \infty} \lim_{x \to \infty} \inf_{n \ge 1} \mathbb{P}\{\cup_{k=0}^{n-1} B_k | M_n > x\} \ge \frac{a}{a+\varepsilon}.$$

Here the left hand term is non-increasing as $\varepsilon \downarrow 0$ while the right hand can be made as close to 1 as we please by choosing a sufficiently small $\varepsilon > 0$. This yields that the limit is equal to 1 for every $\varepsilon > 0$. This completes the proof. □

Now we assume that, in the definition of events B_k, the numbers $A > 0$ and $\varepsilon > 0$ may vary. Namely, we consider a sequence ε_j and a function $h(x) \uparrow \infty$ and introduce disjoint events

$$B_k(x) := \left\{ |S_j + aj| \le j\varepsilon_j + h(x) \text{ for all } j \le k, \ \xi_{k+1} > x + k(a + \varepsilon_k) + h(x) \right\}$$

which have the same intuition behind as events B_k. Let $\varepsilon_j \to 0$ as $j \to \infty$ in such a way that

$$\mathbb{P}\{|S_j + aj| \le j\varepsilon_j \text{ for all } j \ge k\} \to 1 \quad \text{as } k \to \infty;$$

such a sequence exists due to the Strong Law of Large Numbers.

The union of events $B_k(x)$ describes more precisely than Theorem 5.4 the most probable way by which large deviations of M or M_n do occur.

Theorem 5.4*. *Let $\mathbb{E}\xi = -a < 0$ and $F_I \in \mathcal{S}$. Then, for any function $h(x) \to \infty$ such that F_I is h-insensitive,*

$$\mathbb{P}\{\cup_{k=0}^{\infty} B_k(x) | M > x\} \to 1 \quad \text{as } x \to \infty.$$

If, in addition, $F \in \mathcal{S}^*$, then, uniformly in n,

$$\mathbb{P}\{\cup_{k=0}^{n-1} B_k(x) | M_n > x\} \to 1 \quad \text{as } x \to \infty.$$

Proof. Similar arguments as in the proof of Theorem 5.4 may be applied, with some simplifications. □

5.3 Ladder Structure of Maximum of Random Walk

In this section we give an overview of the approach for studying the maximum of a random walk via ascending ladder heights and renewal theory. This approach goes back to Feller [26].

We assume here that the mean of F is negative, so that $S_n \to -\infty$ as $x \to \infty$, with probability 1. Then M is finite with probability 1 and the *first strictly ascending ladder epoch*

$$\eta_+(1) = \eta_+ := \min\{k \ge 1 : S_k > 0\} \le \infty$$

is defective (we put $\min \varnothing = \infty$). The random variable S_{η_+}, which is the first positive sum, is called the *first strictly ascending ladder height*; here $S_\infty = -\infty$. Since the random variables ξ_i are independent and identically distributed and since η_+ is a stopping time, given that $\eta_+ < \infty$, the random variables $\xi_{\eta_+ + k}$, $k = 1, 2, \ldots$, are mutually independent and do not depend on $\{\eta_+, \xi_1, \ldots, \xi_{\eta_+}\}$.

The subsequent (strictly) ascending ladder epochs are defined by induction. If the nth ascending ladder epoch $\eta_+(n)$ is finite, then define the $(n+1)$th ascending ladder epoch $\eta_+(n+1)$ by

$$\eta_+(n+1) := \min\{k > \eta_+(n) : S_k > S_{\eta_+(n)}\} \leq \infty.$$

The random variable $S_{\eta_+(n)}$ is called *nth strictly ascending ladder height*. Again, given that $\eta_+(n) < \infty$, the random variables $\xi_{\eta_+(n)+k}$, $k = 1, 2, \ldots$, are mutually independent and do not depend on $\{\eta_+(1), \ldots, \eta_+(n), \xi_1, \ldots, \xi_{\eta_+(n)}\}$. In particular, if we define

$$p := \mathbb{P}\{M = 0\} = \mathbb{P}\{S_k \leq 0 \text{ for all } k \geq 1\} = \mathbb{P}\{\eta_+(1) = \infty\}, \qquad (5.19)$$

then $\eta_+(n)$ exists with probability $(1-p)^{n-1}$ and it is finite with probability $(1-p)^n$.

The maximum M of the random walk S_n equals the maximal ladder height. Therefore,

$$\mathbb{P}\{M > x\} = \sum_{k=1}^{\infty} \mathbb{P}\{S_{\eta_+(k)} > x, \eta_+(k) < \infty, \eta_+(k+1) = \infty\}$$

$$= p \sum_{k=1}^{\infty} \mathbb{P}\{S_{\eta_+(k)} > x | \eta_+(k) < \infty\}(1-p)^k, \quad x > 0.$$

Let $\{\psi_+(n)\}$ be independent random variables with common distribution

$$G(B) := \mathbb{P}\{\psi_+(n) \in B\} = \mathbb{P}\{S_{\eta_+(1)} \in B | \eta_+(1) < \infty\}.$$

Then, for any Borel set $B \subseteq (0, \infty)$,

$$\mathbb{P}\{M \in B\} = p \sum_{k=1}^{\infty} \mathbb{P}\{\psi_+(1) + \ldots + \psi_+(k) \in B\}(1-p)^k. \qquad (5.20)$$

In other words, the distribution of the maximum M coincides with the distribution of the randomly stopped sum $\psi_+(1) + \ldots + \psi_+(\tau)$, where the stopping time τ is independent of the sequence $\{\psi_+(n)\}$ and is geometrically distributed with parameter $1 - p$, i.e., $\mathbb{P}\{\tau = k\} = p(1-p)^k$ for $k = 0, 1, \ldots$. Equivalently,

$$M =_d G^{*\tau}. \qquad (5.21)$$

5.4 Taboo Renewal Measures

Define the taboo renewal measure on $\mathbb{R}^+$

$$H_+(B) = \mathbb{I}\{0 \in B\} + \sum_{n=1}^{\infty} \mathbb{P}\{S_1 > 0, \ldots, S_n > 0, S_n \in B\}. \qquad (5.22)$$

Since the vector $(\xi_n, \ldots, \xi_1)$ has the same distribution as the vector $(\xi_1, \ldots, \xi_n)$,

$$\mathbb{P}\{S_1 > 0, \ldots, S_n > 0, S_n \in B\}$$
$$= \mathbb{P}\{\xi_n > 0, \xi_n + \xi_{n-1} > 0, \ldots, S_n > 0, S_n \in B\}$$
$$= \mathbb{P}\{S_n - S_{n-1} > 0, S_n - S_{n-2} > 0, \ldots, S_n - S_0 > 0, S_n \in B\}$$
$$= \mathbb{P}\{S_n > S_{n-1}, S_n > S_{n-2}, \ldots, S_n > S_0, S_n \in B\}.$$

The latter event means that S_n is a strict ladder height taking its value in B. Summing over n, we get the following interpretation of the taboo renewal measure H_+. As above, $p = \mathbb{P}\{M = 0\}$.

Lemma 5.5. *For any Borel set $B \subseteq (0, \infty)$, $H_+(B)$ is equal to the mean number of strict ascending ladder heights in the set B. In particular,*

$$H_+(0, \infty) = \sum_{k=1}^{\infty} kp(1-p)^k = (1-p)/p,$$

so that the measure H_+ is finite, and

$$H_+[0, \infty) = 1 + H_+(0, \infty) = 1/p.$$

Let $\eta_-(1) = \eta_- := \min\{k \geq 1 : S_k \leq 0\}$ be the *first (weak) descending ladder epoch* and define subsequent *(weak) descending ladder epochs* by, for $n \geq 1$,

$$\eta_-(n+1) := \min\{k > \eta_-(n) : S_k \leq S_{\eta_-(n)}\}.$$

Define also $\psi_-(1) = \psi_- := S_{\eta_-}$ to be the *first (weak) descending ladder height*, and, for $n \geq 2$, define *n-th descending ladder height* by $\psi_-(n) = S_{\eta_-(n)} - S_{\eta_-(n-1)}$. Since $\mathbb{P}\{S_n \to -\infty\} = 1$, the random variables $\eta_-(n)$ and $\psi_-(n)$ are all proper. Moreover, $\mathbb{E}\eta_- < \infty$ since, by Lemma 5.5,

$$\mathbb{E}\eta_- = 1 + \sum_{n=1}^{\infty} \mathbb{P}\{\eta_- > n\}$$

$$= 1 + \sum_{n=1}^{\infty} \mathbb{P}\{S_1 > 0, \ldots, S_n > 0\}$$

$$= H_+[0, \infty) = 1/p. \tag{5.23}$$

Lemma 5.6 (Wald's identity). *Let τ be a non-negative integer-valued random variable such that, for every n,*

$$\text{the event } \{\tau \leq n\} \text{ does not depend on } \xi_{n+1}. \tag{5.24}$$

Then $\mathbb{E}S_\tau = \mathbb{E}\tau \mathbb{E}\xi_1$, provided both $\mathbb{E}\tau$ and $\mathbb{E}|\xi_1|$ are finite.

Proof. We make use of the following decomposition:

$$\mathbb{E}S_\tau = \sum_{n=0}^{\infty} \mathbb{E}\{S_\tau; \tau = n\} = \sum_{n=1}^{\infty} \sum_{k=1}^{n} \mathbb{E}\{\xi_k; \tau = n\}. \tag{5.25}$$

We may change the order of summation because these double series converge absolutely. Indeed, for non-negative summands,

$$\sum_{n=1}^{\infty} \sum_{k=1}^{n} \mathbb{E}\{|\xi_k|; \tau = n\} = \sum_{k=1}^{\infty} \mathbb{E}\{|\xi_k|; \tau \geq k\}.$$

By condition (5.24), the event $\{\tau \geq k\} = \overline{\{\tau \leq k - 1\}}$ does not depend on ξ_k, so that

$$\sum_{n=1}^{\infty} \sum_{k=1}^{n} \mathbb{E}\{|\xi_k|; \tau = n\} = \sum_{k=1}^{\infty} \mathbb{E}|\xi_k| \mathbb{P}\{\tau \geq k\} = \mathbb{E}\tau \mathbb{E}|\xi_1| < \infty,$$

due to the finiteness of $\mathbb{E}\tau$ and $\mathbb{E}|\xi_1|$. Hence the change of order of summation in (5.25) is justified and we obtain the equality

$$\mathbb{E}S_\tau = \sum_{k=1}^{\infty} \mathbb{E}\{\xi_k; \tau \geq k\}.$$

Again, independence of $\{\tau \geq k\}$ and ξ_k finally yields

$$\mathbb{E}S_\tau = \sum_{k=1}^{\infty} \mathbb{E}\xi_k \mathbb{P}\{\tau \geq k\} = \mathbb{E}\tau \mathbb{E}\xi_1. \qquad \square$$

Applying Wald's identity to the stopping time η_- we get the equality $\mathbb{E}\psi_- = \mathbb{E}\eta_- \mathbb{E}\xi$ which together with (5.23) yields

$$\mathbb{E}\psi_- = -a/p. \tag{5.26}$$

Symmetrically to (5.22), define the taboo renewal measure on $\mathbb{R}^-$

$$H_-(B) = \sum_{n=0}^{\infty} \mathbb{P}\{S_0 \leq 0, \ldots, S_n \leq 0, S_n \in B\}. \tag{5.27}$$

Since $S_n \to -\infty$, $H_-(-\infty, 0] = \infty$. Analogously to the result for H_+, we have the following representation result for the measure H_-, which allows us to deduce useful properties, such as that given by Theorem 5.8

Lemma 5.7. *For any Borel set $B \subseteq (-\infty, 0]$, $H_-(B)$ is equal to the mean number of weak descending ladder heights in the set B, plus 1 if $0 \in B$.*

Theorem 5.8. *Let $\mathbb{E}\xi_1 = -a < 0$. For all $x > 0$,*

$$H_-(-x, 0] \geq \frac{px}{a}.$$

If the distribution F is concentrated on the lattice $\mathbb{Z}$ with the minimal span 1, then

$$H_-\{-k\} \to p/a \quad \text{as } k \to \infty, \ k \in \mathbb{Z}.$$

If the distribution F is non-lattice, then, for any fixed $T > 0$,

$$H_-(-x+[0,T)) \to Tp/a \quad \text{as } x \to \infty.$$

Proof. Define

$$v := \min\{n \geq 1 : S_{\eta_-(n)} \leq -x\}$$
$$= \min\{n \geq 1 : \psi_-(1) + \ldots + \psi_-(n) \leq -x\}.$$

By Lemma 5.7, H_- is the renewal measure generated by the independent identically distributed random variables $\psi_-(n)$, $n \geq 1$. Since v is equal to the number of renewals in the interval $(-x, 0]$ except 0 plus the first one in $(-\infty, -x]$, we have $H(-x, 0] = \mathbb{E}v$.

We have $\psi_-(1) + \ldots + \psi_-(v) \leq -x$. Also, since v is a stopping time, by Wald's identity $\mathbb{E}(\psi_-(1) + \ldots + \psi_-(v)) = \mathbb{E}v\mathbb{E}\psi_-$. Therefore, $\mathbb{E}v \geq -x/\mathbb{E}\psi_-$, where $\mathbb{E}\psi_1^- = -a/p$ from (5.26), so that $\mathbb{E}v \geq px/a$, giving the required proof of the lower bound for $H_-(-x, 0]$.

The local asymptotics for the renewal measure H_- are particular cases of the key renewal theorem, see Feller [26]. □

5.5 Asymptotics for the First Ascending Ladder Height

In this section we again assume that $\mathbb{E}\xi_1 = -a < 0$ so that $S_n \to -\infty$ and define $p := \mathbb{P}\{M = 0\}$.

It follows from the representations (5.20) and (5.21) that the distribution of the maximum M of the random walk $(S_n, n \geq 0)$ is determined by the distribution of the first positive sum S_{η_+} in a rather simple way and we recall in this section a number of results on the latter.

It follows from (5.27) that the distribution of the first ascending ladder height possesses the following representation: for $B \subseteq (0, \infty)$,

$$\mathbb{P}\{S_{\eta_+} \in B\} = \int_{-\infty}^{0} F(B-t)H_-(dt), \tag{5.28}$$

so that

$$\mathbb{P}\{\psi_+ \in B\} = \frac{1}{1-p}\int_{-\infty}^{0} F(B-t)H_-(dt). \tag{5.29}$$

Lemma 5.9. *Under the condition $\mathbb{E}\xi_1 = -a < 0$ and without any further restrictions, the following lower bound holds:*

$$\mathbb{P}\{S_{\eta_+} > x\} \geq \frac{p}{a}\int_x^{\infty} \overline{F}(t)dt \quad \text{for all } x > 0.$$

If, in addition, the integrated tail distribution F_I is long-tailed, then the following tail asymptotics hold:

$$\mathbb{P}\{S_{\eta_+} > x\} \sim p\overline{F_I}(x)/a \quad as\ x \to \infty.$$

Proof. The integral in (5.28) with $B = (x, \infty)$ may be represented as

$$\int_{-\infty}^{0} \overline{F}(x-t)H_-(dt) = (F * H_-)(x, \infty)$$

$$= \int_{0}^{\infty} H_-(-t, 0]F(x+dt). \tag{5.30}$$

Applying here Theorem 5.8 and equality (2.23) we get

$$\int_{0}^{\infty} H_-(-t, 0]F(x+dt) \geq \frac{p}{a} \int_{0}^{\infty} tF(x+dt)$$

$$= \frac{p}{a}\mathbb{E}\{\xi - x; \xi > x\} = \frac{p}{a} \int_{x}^{\infty} \overline{F}(t)dt,$$

which by (5.28) implies the first conclusion of the lemma.

To prove the second assertion of the lemma, fix any $\varepsilon > 0$. It follows from Theorem 5.8 that there exists c such that, for all $t > 0$,

$$H_-(-t, 0] \leq c + (p/a + \varepsilon)t.$$

Then

$$\int_{0}^{\infty} H_-(-t, 0]F(x+dt) \leq \int_{0}^{\infty} (c + (p/a + \varepsilon)t)F(x+dt)$$

$$= c\overline{F}(x) + (p/a + \varepsilon) \int_{x}^{\infty} \overline{F}(t)dt.$$

Since F_I is here assumed long-tailed, by Lemma 2.25, $\overline{F}(x) = o(\overline{F_I}(x))$. Therefore,

$$\int_{0}^{\infty} H_-(-t, 0]F(x+dt) \leq (p/a + 2\varepsilon)\overline{F_I}(x)$$

for all sufficiently large x. Taking now (5.30) and (5.28) with $B = (x, \infty)$ we deduce the upper bound

$$\limsup_{x \to \infty} \frac{\mathbb{P}\{S_{\eta_+} > x\}}{\overline{F_I}(x)} \leq p/a + 2\varepsilon.$$

By letting $\varepsilon \to 0$ and combining this result with the first assertion of the lemma, we complete the proof of the second assertion. $\square$

For local probabilities, the following result holds.

Lemma 5.10. *Let* $\mathbb{E}\xi_1 = -a < 0$ *and let the distribution* F *be long-tailed. If* F *is not lattice then, for any fixed* T,

$$\mathbb{P}\{S_{\eta_+} \in (x, x+T]\} \sim \frac{Tp}{a}\overline{F}(x) \quad \text{as } x \to \infty.$$

If F *is concentrated on the lattice* $\mathbb{Z}$ *with minimal span* 1, *then*

$$\mathbb{P}\{S_{\eta_+} = j\} \sim \frac{p}{a}\overline{F}(j) \quad \text{as } j \to \infty, \ j \in \mathbb{Z}.$$

Proof. Assume that the distribution F is non-lattice (the proof in the lattice case is similar). Then, from the representation (5.28) with $B = (x, x+T]$, we get

$$\mathbb{P}\{S_{\eta_+} \in (x, x+T]\} = \int_{-\infty}^{0} F(x-t, x+T-t]H_-(dt)$$
$$= (F * H_-)(x, x+T]$$
$$= \int_{0}^{\infty} H_-(-t, -t+T]F(x+dt). \quad (5.31)$$

Fix any $\varepsilon > 0$. It follows from Theorem 5.8 that there exists $t_0 > T$ such that

$$Tp/a - \varepsilon \le H_-(-t, -t+T] \le Tp/a + \varepsilon \quad \text{for all } t > t_0.$$

Then, on the one hand, the left inequality implies that

$$\int_{0}^{\infty} H_-(-t, -t+T]F(x+dt) \ge \int_{t_0}^{\infty} (Tp/a - \varepsilon)F(x+dt)$$
$$= (Tp/a - \varepsilon)\overline{F}(x+t_0).$$

On the other hand, the right inequality gives

$$\int_{0}^{\infty} H_-(-t, -t+T]F(x+dt) \le c\int_{0}^{t_0} F(x+dt) + \int_{t_0}^{\infty} (Tp/a + \varepsilon)F(x+dt)$$
$$= cF(x, x+t_0] + (Tp/a + \varepsilon)\overline{F}(x+t_0),$$

where $c := \sup_{t \ge 0} H_-(-t, -t+T]$. Since F is assumed long-tailed, $\overline{F}(x+t_0) \sim \overline{F}(x)$ and $F(x, x+t_0] = \overline{F}(x) - \overline{F}(x+t_0) = o(\overline{F}(x))$ as $x \to \infty$. Hence, for all sufficiently large x,

$$(Tp/a - 2\varepsilon)\overline{F}(x) \le \int_{0}^{\infty} H_-(-t, -t+T]F(x+dt) \le (Tp/a + 2\varepsilon)\overline{F}(x).$$

Substituting these inequalities into (5.31) we obtain that

$$(Tp/a - 2\varepsilon)\overline{F}(x) \le \mathbb{P}\{S_{\eta_+} \in (x, x+T]\} \le (Tp/a + 2\varepsilon)\overline{F}(x).$$

Letting $\varepsilon \to 0$ we complete the proof. $\qquad \square$

Lemma 5.11. *Let* $\mathbb{E}\xi_1 = -a < 0$ *and let the distribution F have a density f on* $\mathbb{R}^+$. *If the function f is long-tailed, then the density of* S_{η_+} *is asymptotically equivalent to* $p\overline{F}(x)/a$ *as* $x \to \infty$.

Proof. By the representation (5.28), the density of S_{η_+} is equal to

$$\int_{-\infty}^0 f(x-t)H_-(dt). \tag{5.32}$$

Since $f(x)$ is long-tailed,

$$F(x,x+1] = \int_x^{x+1} f(y)dy \sim f(x) \quad \text{as } x \to \infty.$$

Thus, the integral (5.32) is asymptotically equivalent to

$$\int_{-\infty}^0 F(x-t,x+1-t]H_-(dt).$$

As shown in the proof of the previous lemma, this integral is asymptotically equivalent to $p\overline{F}(x)/a$ as $x \to \infty$. □

5.6 Tail of the Maximum Revisited

Here we provide an alternative proof of Theorem 5.2 using the ladder height structure of the maximum. Also, we prove the converse result that the standard asymptotics imply subexponentiality.

Theorem 5.12. *Suppose that* $\mathbb{E}\xi = -a < 0$. *Then* $\mathbb{P}\{M > x\} \sim \overline{F}_I(x)/a$ *as* $x \to \infty$ *if and only if* F_I *is subexponential.*

Proof. If F_I is subexponential, then it is in particular long-tailed so that we can apply Lemma 5.9. Then, by Corollary 3.13, the distribution of ψ_+ is subexponential too. From the representation (5.20) and from Theorem 3.37, we thus obtain that, as $x \to \infty$,

$$\mathbb{P}\{M > x\} = p \sum_{k=1}^{\infty} \mathbb{P}\{\psi_+(1) + \ldots + \psi_+(k) > x\}(1-p)^k$$

$$\sim p\mathbb{P}\{\psi_+ > x\} \sum_{k=1}^{\infty} k(1-p)^k$$

$$= \mathbb{P}\{S_{\eta_+} > x\}/p,$$

which, by Lemma 5.9, yields the desired asymptotics for the tail of the maximum.

Now we prove the converse. Assume that $\mathbb{P}\{M > x\} \sim \overline{F}_I(x)/a$ as $x \to \infty$. From the first result of Lemma 5.9 it follows that

$$\mathbb{P}\{\psi_+ > x\} =: \overline{G}(x) \geq \overline{G}_0(x) := \frac{p}{a(1-p)} \int_x^\infty \overline{F}(t)dt \quad \text{for all } x > 0.$$

Since $\overline{G}_0(x) \leq \overline{G}(x)$,

$$p \sum_{k=1}^\infty \overline{G_0^{*k}}(x)(1-p)^k \leq p \sum_{k=1}^\infty \overline{G^{*k}}(x)(1-p)^k$$
$$= \mathbb{P}\{M > x\}$$
$$\sim \overline{F}_I(x)/a \quad \text{as } x \to \infty,$$

by the hypothesis. Applying Theorem 3.38 with geometrically distributed τ to the distribution G_0 on the positive half-line we deduce that G_0 is subexponential. Then, by Corollary 3.13, F_I is subexponential too. □

5.7 Local Probabilities of the Maximum

Any subexponential distribution is long-tailed. Hence, under the conditions of Theorem 5.2, for any fixed T, $\mathbb{P}\{M \in (x, x+T]\} = o(\overline{F}_I(x))$ as $x \to \infty$. Some applications, however, call for more detailed asymptotics of the random walk maximum than are given by Theorem 5.2. For this, we need F to be strong subexponential, $F \in \mathcal{S}^*$, which, by Theorem 3.27, implies that $F_I \in \mathcal{S}$.

As above, for $T \in (0, \infty)$, we put $\Delta = (0, T]$.

Theorem 5.13. *Suppose that* $\mathbb{E}\xi = -a < 0$ *and that the distribution F is long-tailed. Then the following statements are equivalent:*
(i) $F \in \mathcal{S}^*$.
(ii) $\mathbb{P}\{M \in (x, x+T]\} \sim T\overline{F}(x)/a$ *as* $x \to \infty$
    (if the distribution F is lattice, then x and T should be restricted to values of the lattice span).

Proof. Assume that the distribution F is non-lattice (the proof in the lattice case is similar). Since $F \in \mathcal{L}$ we have from Lemma 5.10 that

$$\mathbb{P}\{S_{\eta_+} \in (x, x+T]\} \sim \frac{Tp}{a}\overline{F}(x) \quad \text{as } x \to \infty. \tag{5.33}$$

Again since $F \in \mathcal{L}$ it follows in particular that the distribution of ψ_+ is Δ-long-tailed. It follows from Theorem 4.32 (and the result that membership of $\mathcal{S}^*$ is a tail property of a distribution) that the relation $F \in \mathcal{S}^*$ is equivalent to the Δ-subexponentiality of the integrated tail distribution F_I. In turn, it follows from the equivalence (5.33) and from Corollary 4.24 that the distribution of ψ_+ is Δ-subexponential if and only if F_I is so. Thus, the distribution of ψ_+ is Δ-subexponential if and only if $F \in \mathcal{S}^*$.

From the representation (5.20) and from Theorem 3.37, we have

$$\mathbb{P}\{M \in x + \Delta\} = p \sum_{k=1}^{\infty} \mathbb{P}\{\psi_+(1) + \ldots + \psi_+(k) \in x + \Delta\}(1-p)^k.$$

By Theorem 4.31, the probability on the right is asymptotically equivalent to

$$\mathbb{P}\{\psi_+ \in x + \Delta\} p \sum_{k=1}^{\infty} k(1-p)^k = \mathbb{P}\{\psi_+ \in x + \Delta\}(1-p)/p$$

as $x \to \infty$ if and only if the distribution of ψ_+ is Δ-subexponential. By (5.33),

$$\mathbb{P}\{\psi_+ \in x + \Delta\}(1-p)/p = \mathbb{P}\{S_{\eta_+} \in x + \Delta\}/p$$
$$\sim T\overline{F}(x)/a \quad \text{as } x \to \infty.$$

Combining the above statements, we thus obtain the stated result.   □

To conclude this section note that it follows from the above theorem that if $F \in S^*$ then, for any $T > 0$,

$$\mathbb{P}\{M \in (x, x+T]\} \sim \frac{1}{a} \int_x^{x+T} \overline{F}(y) dy \quad \text{as } x \to \infty.$$

It therefore follows that, for any $T_0 > 0$, this asymptotic result holds uniformly in $T > T_0$.

5.8 Density of the Maximum

The distribution of the maximum M of a random walk with negative drift always contains an atom at the origin, i.e. $\mathbb{P}\{M = 0\} = 1 - p$. In this section we are concerned with the absolutely continuous part of the distribution of M. We assume that the conditional distribution of ξ given that $\xi > 0$ is absolutely continuous. Thus let $f(x) \geq 0$ be a function such that

$$F(B) = \int_B f(x) dx \quad \text{for any } B \in \mathcal{B}(0, \infty).$$

Then the distribution of the maximum M is the sum of the atomic distribution of weight p at zero and an absolutely continuous distribution with a density $m(x)$, where $\int_0^\infty m(x) dx = 1 - p$.

Theorem 5.14. *Suppose that* $\mathbb{E}\xi = -a < 0$ *and that* f *is long-tailed. Then the following statements are equivalent:*
(i) $F \in S^*$.
(ii) $m(x) \sim \overline{F}(x)/a$ *as* $x \to \infty$.

Proof. Since f is long-tailed we can apply Lemma 5.11 which states that the density $g(x)$ of S_{η_+} is asymptotically equivalent to $p\overline{F}(x)/a$ as $x \to \infty$. In particular, the density $g(x)/(1-p)$ of ψ_+ is long-tailed. By Theorem 4.32, (and again the result that membership of $\mathcal{S}^*$ is a tail property of a distribution) the relation $F \in \mathcal{S}^*$ is equivalent to subexponentiality of the integrated tail density of F_I. In turn, it follows from the equivalence $g(x) \sim p\overline{F}(x)/a$ and from Theorem 4.8 that the density of ψ_+ is subexponential if and only if the density of F_I is also. Thus, the density of ψ_+ is subexponential if and only if $F \in \mathcal{S}^*$.

From the representation (5.20) and from Theorem 3.37, we obtain

$$m(x) = p \sum_{k=1}^{\infty} (g/(1-p))^{*k}(x)(1-p)^k.$$

By Theorem 4.30, the density on the right is asymptotically equivalent to

$$\frac{g(x)}{1-p} p \sum_{k=1}^{\infty} k(1-p)^k = \frac{g(x)}{p} \sim \overline{F}(x)$$

as $x \to \infty$ if and only if the density of ψ_+ is subexponential. The above statements together yield the desired equivalence of (i) and (ii). $\qquad\square$

5.9 Explicitly Calculable Ascending Ladder Heights

There are only a few cases where the distribution of the first strictly ascending ladder height S_{η_+} may be explicitly calculated. This leads to new local theorems and to better bounds and asymptotics than are in general available. We consider three such cases, one relates to exponential left tail distributions and two others to lattice distributions.

Exponential Case

We start with the case, where the distribution function $\mathbb{P}\{\xi \le x\}$ is exponential for negative x. This case is important in applications (e.g. in risk theory (see Sect. 5.11) and in queueing (see Sect. 5.10)) where ξ may be represented as a difference $\sigma - \tau$ of two independent random variables and τ has an exponential distribution.

We thus assume that the left tail of the distribution of ξ is exponential, i.e., for some c and $\alpha > 0$, $\mathbb{P}\{\xi \le x\} = ce^{\alpha x}$ for all $x \le 0$. Then the distribution of the first weakly descending ladder height $\psi_- = S_{\eta_-}$ is given by, for $x \le 0$,

$$\mathbb{P}\{\psi_- \le x\} = \int_0^\infty F(x-t)H_+(dt)$$

$$= ce^{\alpha x} \int_0^\infty e^{-\alpha t} H_+(dt),$$

so that $c \int_0^\infty e^{-\alpha t} H_+(dt)$ must be equal to one, and ψ_- is exponentially distributed with parameter α. Hence, the sequence $\psi_-(n)$ generates the Poisson process with intensity α. Therefore, by Lemma 5.7, H_- is the sum of an α multiple of Lebesgue measure plus a unit mass at 0. Then the equality (5.28) yields that, for any Borel set $B \subset (0,\infty)$,

$$\mathbb{P}\{S_{\eta_+} \in B\} = \alpha \int_{-\infty}^0 F(B-t)dt + F(B),$$

so that, for $x > 0$,

$$\mathbb{P}\{S_{\eta_+} > x\} = \alpha \int_x^\infty \overline{F}(t)dt + \overline{F}(x). \tag{5.34}$$

In addition, by (5.26),

$$\mathbb{P}\{M = 0\} = \mathbb{P}\{\eta_+ = \infty\} = \mathbb{E}\xi/\mathbb{E}S_{\eta_-} = -\mathbb{E}\xi\,\alpha. \tag{5.35}$$

The case where $\xi = \sigma - \tau$, where σ and τ are independent and non-negative, is of special interest as it arises naturally in both queueing and risk theory. If τ is exponentially distributed with parameter $\alpha > 0$ then, for $x \le 0$,

$$\mathbb{P}\{\xi \le x\} = \int_0^\infty e^{\alpha(x-y)}\mathbb{P}\{\sigma \in dy\}$$

$$= e^{\alpha x} \int_0^\infty e^{-\alpha y} B(dy),$$

where B is the distribution of σ, so that $\mathbb{P}\{\xi \le x\}$ is an exponential function for $x \le 0$. In addition, the distribution F is absolutely continuous with density f, say, and for $x > 0$

$$f(x) = \alpha \int_x^\infty e^{-\alpha(y-x)} B(dy),$$

$$\overline{F}(x) = \int_x^\infty (1 - e^{-\alpha(y-x)}) B(dy) = \overline{B}(x) - f(x)/\alpha.$$

Hence, by (5.34) the density of S_{η_+} is equal to

$$\alpha \overline{F}(x) + f(x) = \alpha \overline{B}(x).$$

Finally, by (5.35)

$$\mathbb{P}\{M = 0\} = (1/\alpha - \mathbb{E}\sigma)\alpha = 1 - \rho,$$

where $\rho := \mathbb{E}\sigma/\mathbb{E}\tau = \alpha\mathbb{E}\sigma$. Hence the representation (5.20) for the tail distribution of the maximum M of the random walk simplifies to:

$$\mathbb{P}\{M > x\} = (1-\rho) \sum_{n=0}^\infty \rho^n \overline{B_r^{*n}}(x) \tag{5.36}$$

where B_r is a proper distribution on $\mathbb{R}^+$ with density $\overline{B}(x)/\mathbb{E}\sigma$.

Applying here the inequality $B_r^{*n}(x) \leq B_r^n(x)$ we obtain the lower bound

$$\mathbb{P}\{M > x\} \geq (1-\rho) \sum_{n=0}^{\infty} \rho^n (1 - B_r^n(x))$$

$$= 1 - \frac{1-\rho}{1-\rho+\rho\overline{B}_r(x)} = \frac{\overline{B}_r(x)}{\frac{1-\rho}{\rho} + \overline{B}_r(x)}. \tag{5.37}$$

This is slightly better than the general lower bound delivered by Theorem 5.1, since the function $\frac{t}{a+t}$ increases in t and $\overline{B}(x) \geq \overline{F}(x)$, so that $\overline{B}_r(x) \geq \overline{F}_I(x)/\mathbb{E}\sigma$.

Some results proved earlier for the distribution of the maximum M become easier in this case. In particular this distribution of M is the sum of an atom of mass $1 - \rho$ at zero and of an absolutely continuous part on $(0, \infty)$ with tail function given, for $x > 0$, by

$$(1-\rho) \sum_{n=1}^{\infty} \rho^n \overline{B}_r^{*n}(x). \tag{5.38}$$

Consider the monotonically decreasing density $b(x) := \overline{B}(x)/\mathbb{E}\sigma$. Then the density $m(x)$ of the absolutely continuous part of the distribution of M is given by

$$m(x) := (1-\rho) \sum_{n=1}^{\infty} \rho^n b^{*n}(x)$$

Further Theorem 5.14 may be simplified as follows.

Theorem 5.15. *Suppose that $\xi = \sigma - \tau$, that τ has exponential distribution with parameter $\alpha > 0$, that $\sigma \geq 0$ has distribution B, that σ and τ are independent and that $\rho := \alpha\mathbb{E}\sigma < 1$. Then the following statements are equivalent:*
  (i) $B \in \mathcal{S}^*$.
  (ii) $m(x) \sim \frac{\alpha}{1-\rho}\overline{B}(x)$ *as* $x \to \infty$.

Geometric Case

This is a lattice analogue of the previous case. Suppose that ξ takes values in $\mathbb{Z}$ and that the left tail of ξ is geometrically distributed, i.e., for some $c > 0$ and $q \in (0,1)$, $\mathbb{P}\{\xi = -n\} = cq^n$ for all $n = 0, 1, \ldots$. Then

$$\mathbb{P}\{\psi_- = -n\} = \sum_{k=0}^{\infty} \mathbb{P}\{\xi = -n-k\}H_+\{k\}$$

$$= cq^n \sum_{k=0}^{\infty} q^k H_+\{k\},$$

so ψ_- is geometrically distributed with parameter q, i.e., $\mathbb{P}\{\psi_- = -n\} = (1-q)q^n$, $n = 0, 1, \ldots$. Hence, the mean number of weak descending ladder heights $\psi_-(n)$

at any point of $\mathbb{Z}^-$ equals $(1-q)/q$. Therefore, by Lemma 5.7, H_- is the sum of a $(1-q)/q$ multiple of the counting measure on $\mathbb{Z}^-$ plus a unit mass at 0. Then the equality (5.28) yields, for any $n \geq 1$,

$$
\begin{aligned}
\mathbb{P}\{S_{\eta_+} = n\} &= \sum_{k=0}^{\infty} F\{n+k\}H_-\{-k\} \\
&= \frac{1-q}{q} \sum_{k=0}^{\infty} F\{n+k\} + F\{n\} = \frac{1-q}{q}F[n,\infty) + F\{n\}. \quad (5.39)
\end{aligned}
$$

In addition, by (5.26),

$$
\mathbb{P}\{M = 0\} = \mathbb{P}\{\eta_+ = \infty\} = \mathbb{E}\xi/\mathbb{E}S_{\eta_-} = -\mathbb{E}\xi(1-q)/q. \quad (5.40)
$$

Consider the case where $\xi = \sigma - \tau$, where σ and τ are independent and both take values in $\mathbb{Z}^+$. If τ is geometrically distributed with parameter $q \in (0,1)$ then, for $n \geq 0$,

$$
\begin{aligned}
\mathbb{P}\{\xi = -n\} &= \sum_{k=0}^{\infty}(1-q)q^{-n-k}\mathbb{P}\{\sigma = k\} \\
&= (1-q)q^{-n}\sum_{k=0}^{\infty}q^{-k}B\{k\},
\end{aligned}
$$

where B is the distribution of σ. So, we are in the framework of geometrically distributed left tail of ξ. In addition, for $n > 0$,

$$
F\{n\} = (1-q)\sum_{k=n}^{\infty} q^{k-n}B\{k\},
$$

$$
F[n,\infty) = \sum_{k=n}^{\infty}(1-q^{k-n+1})B\{k\} = B[n,\infty) - \frac{q}{1-q}F\{n\}.
$$

Hence, by (5.39),

$$
\mathbb{P}\{S_{\eta_+} = n\} = \frac{1-q}{q}B[n,\infty).
$$

Finally, by (5.40),

$$
\mathbb{P}\{M = 0\} = (q/(1-q) - \mathbb{E}\sigma)(1-q)/q = 1 - \rho,
$$

where $\rho := \mathbb{E}\sigma/\mathbb{E}\tau = (1-q)\mathbb{E}\sigma/q$. So, the representation (5.20) for the distribution of the maximum M of the random walk simplifies to:

$$
\mathbb{P}\{M = n\} = (1-\rho)\sum_{k=0}^{\infty}\rho^k B_r^{*k}\{n\}, \quad (5.41)
$$

where B_r is a proper distribution on $\{1,2,\dots\}$ with probabilities

$$
B_r\{n\} := B[n,\infty)/\mathbb{E}\sigma.
$$

Thus, we have the following analogue of Theorem 5.15.

Theorem 5.16. *Suppose that $\xi = \sigma - \tau$, with τ having geometric distribution with parameter $q \in (0,1)$, $\sigma \geq 0$ having distribution B, σ, and τ are independent, and that $\mathbb{E}\sigma < q/(1-q)$. Then the following statements are equivalent:*
  (i) $B \in \mathcal{S}^*$.
  (ii) $\mathbb{P}\{M = n\} \sim \frac{1}{\mathbb{E}\tau - \mathbb{E}\sigma}B[n,\infty)$ *as $n \to \infty$.*

Left Continuous Random Walk

Here we consider another type of a lattice distribution F. Suppose that a random walk takes values in $\mathbb{Z}$ and that $F(-\infty, -2] = \mathbb{P}\{\xi \leq -2\} = 0$. Then it is called a *left continuous* or a *left skip-free* random walk. Then the first weakly descending ladder height $\psi_- = S_{\eta_-}$ takes two values only, -1 and 0, with probabilities

$$\mathbb{P}\{\psi_- = -1\} = \mathbb{P}\{\xi_1 = -1\} =: p_{-1};$$
$$\mathbb{P}\{\psi_- = 0\} = 1 - p_{-1}.$$

Therefore, H_- is a $1/p_{-1}$ multiple of the counting measure on $\mathbb{Z}^-$. Then the equality (5.28) yields, for any $n \geq 1$,

$$\mathbb{P}\{S_{\eta_+} = n\} = \sum_{k=0}^{\infty} F\{n+k\}H_-\{-k\} = \frac{1}{p_{-1}}F[n,\infty).$$

In addition, by (5.26),

$$p := \mathbb{P}\{M = 0\} = \mathbb{P}\{\eta_+ = \infty\} = \mathbb{E}\xi/\mathbb{E}S_{\eta_-} = -\mathbb{E}\xi/p_{-1}$$

and

$$1 - p = (p_{-1} + \mathbb{E}\xi)/p_{-1} = \mathbb{E}\{\xi; \xi \geq 1\}/p_{-1}.$$

Hence, in the case of left continuous random walk

$$\mathbb{P}\{\psi_+ = n\} = \frac{\mathbb{P}\{S_{\eta_+} = n\}}{1-p} = \frac{F[n,\infty)}{\mathbb{E}\{\xi; \xi \geq 1\}}$$

and the representation (5.20) for the distribution of the maximum M of the random walk simplifies to:

$$\mathbb{P}\{M = n\} = p\sum_{k=0}^{\infty}(1-p)^k F_0^{*k}\{n\}, \tag{5.42}$$

where F_0 is a proper distribution on $\{1,2,\dots\}$ with probabilities

$$F_0\{n\} := F[n,\infty)/\mathbb{E}\{\xi; \xi \geq 1\}.$$

5.10 Single Server Queueing System

We consider the basic model in queueing theory, a *first-come-first-served* (FCFS) single server system, $GI/GI/1$, where the customers are served in the order of arrival by a single server. The dynamics of a single server system are described as follows. Customers arrive one at a time into the system with interarrival time τ_n between $(n-1)$th and nth successive customers. Customers form a queue and the first customer in the queue moves immediately to a server when the server becomes idle. The service time of the nth customer is denoted by σ_n. After being served the customer leaves the system.

Let τ be a typical interarrival time and σ a typical service time. Independent identically distributed sequences of interarrival times $\{\tau_n\}$ with finite mean and service times $\{\sigma_n\}$ with finite mean are assumed to be mutually independent.

For $n = 1, 2, \ldots$, let W_n be waiting time of the nth customer, which is the residual workload in the queue observed by the nth customer upon its arrival into the system (or the delay which customer n experiences). Then the waiting times W_n satisfy the Lindley recursion [41]:

$$W_{n+1} = (W_n + \sigma_n - \tau_{n+1})^+. \tag{5.43}$$

In the setting of independence of jumps, W_n is a Markov chain which is called a *random walk with delay at the origin*.

We assume that the system is *stable*, i.e. $\rho := \mathbb{E}\sigma/\mathbb{E}\tau < 1$. Then the Markov chain W_n has a unique stationary distribution, and the distribution of W_n converges as $n \to \infty$ to the stationary distribution in the total variation norm. This follows from the strong law of large numbers and from the following result.

Lemma 5.17. *Given $W_1 = 0$, W_{n+1} coincides in distribution with $M_n := \max(S_k, k \le n)$ where $S_0 = 0$ and $S_n = \sum_{k=1}^{n}(\sigma_k - \tau_{k+1})$ for $n \ge 1$.*

Proof. Denote $\xi_n := \sigma_n - \tau_{n+1}$. It follows from the recursion (5.43) that

$$\begin{aligned}
W_{n+1} &= \max(0, W_n + \xi_n) \\
&= \max(0, \xi_n, W_{n-1} + \xi_{n-1} + \xi_n) \\
&\quad \cdots \\
&= \max(0, \xi_n, \xi_{n-1} + \xi_n, \xi_{n-2} + \xi_{n-1} + \xi_n, \ldots, \xi_1 + \cdots + \xi_n).
\end{aligned}$$

Now the result follows because the vector $(\xi_n, \ldots, \xi_1)$ has the same distribution as the vector $(\xi_1, \ldots, \xi_n)$.   □

Note that in the above lemma the finite dimensional distributions of the processes $\{W_n, n \ge 1\}$ and $\{M_n, n \ge 1\}$ are different. To see this, it is enough to observe that the trajectories of $\{M_n, n \ge 1\}$ are non-decreasing while those of $\{W_n, n \ge 1\}$ are clearly not.

Assume that the distribution of σ has unbounded support, i.e. $B(x) := \mathbb{P}\{\sigma \leq x\} < 1$ for all x. The first result is about the tail behaviour of the stationary waiting time distribution in the case of a heavy-tailed distribution for σ.

Theorem 5.18. *Let W be a random variable distributed as the stationary waiting time. If the residual service time distribution B_r is subexponential, then*

$$\mathbb{P}\{W > x\} \sim \frac{\rho}{1-\rho}\overline{B}_r(x) \quad as\ x \to \infty. \tag{5.44}$$

If, in addition, $B \in \mathcal{S}^*$ then, for every $T > 0$,

$$\mathbb{P}\{x < W \leq x + T\} \sim \frac{T}{\mathbb{E}\tau - \mathbb{E}\sigma}\overline{B}(x) \quad as\ x \to \infty; \tag{5.45}$$

in the lattice case x and T should be restricted to values of the lattice span.

Proof. Let F be the distribution of the difference $\sigma - \tau$. Since

$$F(x) = \int_0^\infty \overline{B}(x+y)\mathbb{P}\{\tau \in dy\},$$

we have, eventually in x,

$$\overline{F_I}(x) = \mathbb{E}\sigma \int_0^\infty \overline{B}_r(x+y)\mathbb{P}\{\tau \in dy\}.$$

If B_r is subexponential then, by Corollary 3.18, F_I is subexponential too and $\overline{F_I}(x) \sim \overline{B}_r(x)$ as $x \to \infty$. Thus the equivalence (5.44) follows from Theorem 5.12.

If $B \in \mathcal{S}^*$, then F is strong subexponential too and $\overline{F}(x) \sim \overline{B}(x)$ as $x \to \infty$, so that the equivalence (5.45) follows from Theorem 5.13. $\square$

Now consider a simple $M/GI/1$ system with Poisson arrival process of intensity α and with general service times σ_n. Here τ's are exponentially distributed with mean $\mathbb{E}\tau = 1/\alpha$. Making use of explicit formulas for the exponential case in Sect. 5.9 we may specify the tail behaviour of the stationary waiting time in the following way.

Theorem 5.19. *In an $M/GI/1$ system, the following are equivalent:*

(i) The residual service time distribution B_r is subexponential.
(ii) $\mathbb{P}\{W > x\} \sim \frac{\rho}{1-\rho}\overline{B}_r(x)$ as $x \to \infty$.
    Also, the following are equivalent:
(iii) The service time distribution B is strong subexponential.
(iv) The density $m(x)$ of the absolutely continuous part of the distribution of W is equivalent to $\frac{\alpha}{1-\rho}\overline{B}(x)$ as $x \to \infty$.

Proof. By the previous theorem, (i) implies (ii). The converse, i.e., the implication (ii)$\Rightarrow$(i) follows from the representation (5.38) and Theorem 3.38 with geometrically distributed stopping time.

The equivalence of (iii) and (iv) was proved in Theorem 5.15. $\square$

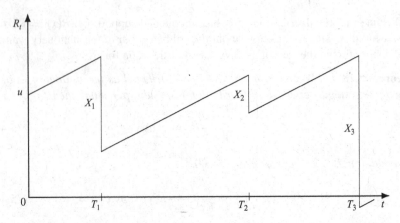

Fig. 5.1 A typical trajectory leading to ruin

5.11 Ruin Probabilities in Cramér–Lundberg Model

In context of the collective theory of risk, we consider the classical *Cramér–Lundberg model* (or the *compound Poisson model*) defined as follows. We consider an insurance company and assume the constant inflow of premium occurs at rate c, i.e., the premium income is assumed to be linear in time with rate c. Also assume that the claims incurred by the insurance company arrive according to a homogeneous Poisson process N_t with intensity λ and the sizes (amounts) $X_n \geq 0$ of the claims are independent identically distributed random variables with common distribution B and mean b. The X's are assumed to be independent of the process N_t. The company has an initial risk reserve $u = R_0 \geq 0$.

Then the risk reserve R_t at time t is equal to

$$R_t = u + ct - \sum_{i=1}^{N_t} X_i.$$

The probability

$$\mathbb{P}\{R_t \geq 0 \text{ for all } t \geq 0\} = \mathbb{P}\left\{\min_{t \geq 0} R_t \geq 0\right\}$$

is the probability of ultimate survival and

$$\psi(u) := \mathbb{P}\{R_t < 0 \text{ for some } t \geq 0\}$$
$$= \mathbb{P}\left\{\min_{t \geq 0} R_t < 0\right\}$$

is the probability of ruin. The techniques developed for random walks provide a method for estimating the probability of ruin in the presence of heavy-tailed distribution for claim sizes. We have

$$\psi(u) = \mathbb{P}\left\{\sum_{i=1}^{N_t} X_i - ct > u \text{ for some } t \geq 0\right\}.$$

Since $c > 0$, the ruin can only occur at a claim epoch, see Fig. 5.1. Therefore,

$$\psi(u) = \mathbb{P}\Big\{ \sum_{i=1}^{n} X_i - cT_n > u \text{ for some } n \geq 1 \Big\},$$

where T_n is the nth claim epoch, so that $T_n = \tau_1 + \ldots + \tau_n$ where the τ's are independent random variables with common exponential distribution with parameter λ. Denote $\xi_i := X_i - c\tau_i$ and $S_n := \xi_1 + \ldots + \xi_n$, then

$$\psi(u) = \mathbb{P}\Big\{ \sup_{n \geq 1} S_n > u \Big\}.$$

This relation represents the ruin probability problem as the tail probability problem for the maximum of the associated random walk S_n. Let the *net-profit condition* $c > b\lambda$ hold, thus S_n has a negative drift and $\psi(u) \to 0$ as $u \to \infty$. Similar to Theorem 5.19, we deduce the following result on the decreasing rate of the ruin probability to zero as the initial risk reserve becomes large in the case of heavy-tailed claim size distribution.

Theorem 5.20. *In the compound Poisson risk model, let $c > b\lambda$. Then the following are equivalent:*

 (i) The integrated tail claim size distribution B_I is subexponential.
(ii) $\psi(u) \sim \frac{\lambda}{c-b\lambda} \overline{B}_I(u)$ as $u \to \infty$.
  Also, the following are equivalent:
(iii) The claim size distribution B is strong subexponential.
(iv) $\psi'(u) \sim -\frac{\lambda}{c-b\lambda} \overline{B}(u)$ as $u \to \infty$.

For the compound Poisson risk model, the inequality (5.37) may be rewritten as follows:

$$\psi(u) \geq \frac{\int_u^\infty \overline{B}(v)dv}{c/\lambda - b + \int_u^\infty \overline{B}(v)dv}.$$

This allows us to determine the lower bound for the initial risk reserve for an insurance company to set the desired level of the ruin probability.

The next result deals with the finite time horizon probability of ruin, i.e., with

$$\psi(u,t) := \mathbb{P}\{R_s < 0 \text{ for some } s \in [0,t]\}$$
$$= \mathbb{P}\Big\{ \sum_{i=1}^{n} X_i - cT_n > u \text{ for some } n \leq N_t \Big\}.$$

Theorem 5.21. *In the compound Poisson risk model, let $c > b\lambda$. If the claim size distribution B is strong subexponential, then, uniformly in $t \geq 0$,*

$$\psi(u,t) \sim \frac{\lambda}{c-b\lambda} \int_u^{u+t(c-b\lambda)} \overline{B}(v)dv \quad as \ u \to \infty.$$

Proof. Since $B \in \mathcal{S}^*$, B is subexponential. Hence, by Theorem 3.37 with stopping time having Poisson distribution with parameter λt, for every fixed t,

$$\mathbb{P}\left\{\sum_{i:T_i \le t} X_i > u\right\} = \sum_{j=0}^{\infty} \frac{(\lambda t)^j}{j!} e^{-\lambda t} \overline{B^{*j}}(u)$$

$$\sim \lambda t \overline{B}(u) \quad \text{as } u \to \infty, \tag{5.46}$$

and this equivalence holds uniformly in t on any compact set. This observation together with bounds

$$\mathbb{P}\left\{\sum_{i:T_i \le t} X_i > u + ct\right\} \le \psi(u,t) \le \mathbb{P}\left\{\sum_{i:T_i \le t} X_i > u\right\}$$

and long-tailedness of B yield that, uniformly in t on any compact set,

$$\psi(u,t) \sim \lambda t \overline{B}(u) \quad \text{as } u \to \infty.$$

Taking into account that

$$\lambda t \overline{B}(u + t(c - b\lambda)) \le \frac{\lambda}{c - b\lambda} \int_u^{u+t(c-b\lambda)} \overline{B}(v)dv \le \lambda t \overline{B}(u)$$

and that B is long-tailed, we prove the result of the theorem for t on any compact set. Therefore, it remains to prove that

$$\psi(u,t) \sim \frac{\lambda}{c - b\lambda} \int_u^{u+t(c-b\lambda)} \overline{B}(v)dv \quad \text{as } u,t \to \infty. \tag{5.47}$$

Since B is long-tailed, it is sufficient to prove the latter equivalence for t going to infinity along integers.

For $k = 1, 2, \ldots$, put

$$\eta_k := R_{k-1} - R_k = \sum_{i:T_i \in (k-1,k]} X_i - c$$

and

$$\hat{\eta}_k := \max_{s \in (0,1]} (R_{k-1} - R_{k-1+s}) = \max_{s \in (0,1]} \left(\sum_{i:T_i \in (k-1,k-1+s]} X_i - cs\right).$$

Then the tail probability $\psi(u,t)$ for integer t may be represented as follows:

$$\psi(u,t) = \mathbb{P}\left\{\max_{k \le t-1} \left(\sum_{i=1}^k \eta_i + \hat{\eta}_{k+1}\right) > u\right\}.$$

Since the X's are non-negative, $\hat{\eta}_k \le \eta_k + c$. By the definition of η and properties of the Poisson distribution, η's are independent identically distributed random variables, and $\mathbb{E}\eta = \lambda b - c$. By (5.46), $\mathbb{P}\{\eta > u\} \sim \lambda \overline{B}(u)$ as $u \to \infty$. Thus,

$$\psi(u,t) \leq \mathbb{P}\left\{\max_{k \leq t} \sum_{i=1}^{k} \eta_i > u - c\right\}$$

$$\sim \frac{\lambda}{c - b\lambda} \int_{u-c}^{u-c+t(c-b\lambda)} \overline{B}(v)dv$$

$$\sim \frac{\lambda}{c - b\lambda} \int_{u}^{u+t(c-b\lambda)} \overline{B}(v)dv \quad \text{as } u \to \infty,$$

by strong subexponentiality of B and Theorem 5.3. On the other hand, $\hat{\eta}_k \geq \eta_k$, so that

$$\psi(u,t) \geq \mathbb{P}\left\{\max_{k \leq t} \sum_{i=1}^{k} \eta_i > u\right\}$$

$$\sim \frac{\lambda}{c - b\lambda} \int_{u}^{u+t(c-b\lambda)} \overline{B}(v)dv \quad \text{as } u \to \infty.$$

Altogether implies (5.47) and the proof is complete. $\qquad\square$

5.12 Subcritical Branching Processes

The concept of a subexponential distribution was introduced by Chistyakov in [13] in 1964, in the context of branching processes. The book by Athreya and Ney [7] on branching processes published in 1972 was the first book where subexponential distributions have been regularly introduced. In this section we pay tribute to the origins of the theory of subexponential distributions.

An *age-dependent branching process* is a stochastic process $\{X_t\}$ valued in $\mathbb{Z}^+$ representing the number of particles existing at time $t \in \mathbb{R}^+$. Here the particles have general lifetime distribution G on $[0, \infty)$. At time $t = 0$ there is exactly one parent particle which lives for time T_0 having distribution G, and then splits independently of everything else into a random number $\xi^{(1)}$ of offspring according to the probability distribution F on $\mathbb{Z}^+$. It constitutes the first generation of $X_1 = \xi^{(1)}$ particles. These live for times $T_1^{(1)}, \ldots, T_{X_1}^{(1)}$, and then split into $\xi_1^{(2)}, \ldots, \xi_{X_1}^{(2)}$ of offspring according to the distribution F. All primitive random variables are assumed to be mutually independent. The number of particles in consequent generations is linked by the recursive formula

$$X_{n+1} = \sum_{j=1}^{X_n} \xi_j^{(n+1)},$$

where $\{\xi_j^{(n)}\}$ is a family of independent identically distributed non-negative integer-valued random variables with common distribution F. Let $X_t^{(k)}$ be independent copies of the process X_t. By the Markov property, for every t, we obtain the following equality in distribution

$$X_t \overset{d}{=} \mathbb{I}\{T_0 > t\} + \mathbb{I}\{T_0 \le t\} \sum_{k=1}^{\xi^{(1)}} X_{t-T_0}^{(k)}. \tag{5.48}$$

Let $m := \mathbb{E}\xi^{(1)} = f'(1)$ be finite. The branching process is called *subcritical* if $m < 1$; *critical* if $m = 1$; and *supercritical* if $m > 1$. Hereinafter in this section we deal with a subcritical branching process. Then (see e.g., [7, Theorem 1.5.1]) the extinction probability of X_t, $\mathbb{P}\{X_t = 0 \text{ for some } t\}$, equals 1. The next result specifies the rate of extinction for a subcritical branching process with heavy-tailed lifetime distribution G.

Theorem 5.22. *Let $m < 1$. Then, for every $t > 0$,*

$$\mathbb{E}X_t \ge \mathbb{P}\{X_t \ge 1\} \ge \frac{\overline{G}(t)}{1 - mG(t)}. \tag{5.49}$$

If, in addition, the lifetime distribution G is subexponential, then

$$\mathbb{E}X_t \sim \mathbb{P}\{X_t \ge 1\} \sim \mathbb{P}\{X_t = 1\} \sim \frac{\overline{G}(t)}{1 - m} \qquad \text{as } t \to \infty. \tag{5.50}$$

In particular, $\mathbb{P}\{X_t = 1 | X_t \ge 1\} \to 1$ as $t \to \infty$.

Proof. Taking expectation in (5.48), we obtain the following equation for $\mathbb{E}X_t$:

$$\mathbb{E}X_t = \overline{G}(t) + m \int_0^t \mathbb{E}X_{t-y} G(dy), \tag{5.51}$$

by Wald's identity. In the case $m < 1$, this equation implies $\mathbb{E}X_t \le 1$ for every t. Indeed, if $G(t) = 0$, then $X_s = 1$ for all $s \le t$. Otherwise we find $t_0 \le t$ such that $G(t_0) > 0$ and

$$\mathbb{E}X_{t_0} \ge (\overline{G}(t_0) + mG(t_0)) \sup_{s \le t} \mathbb{E}X_s,$$

which is possible because $\overline{G}(t_0) + mG(t_0) < 1$. Then it follows from (5.51) that

$$\mathbb{E}X_{t_0} \le \overline{G}(t_0) + m \sup_{s \le t_0} \mathbb{E}X_s G(t_0) \le \overline{G}(t_0) + m \sup_{s \le t} \mathbb{E}X_s G(t_0),$$

which together with the previous inequality yields

$$\sup_{s \le t} \mathbb{E}X_s \le 1. \tag{5.52}$$

Further, the condition $m < 1$ also yields that the function $\mathbb{E}X_t$ is non-increasing. Indeed, let $s < t$ and $\tau_1, \ldots, \tau_{X_s}$ be the time epochs of splitting for the particles living at time s. Make use of the decomposition

$$X_t \stackrel{d}{=} \sum_{k=1}^{X_s} \mathbb{I}\{\tau_k > t\} + \sum_{k=1}^{X_s} \mathbb{I}\{\tau_k \leq t\} \sum_{i=1}^{\xi(k)} X_{t-\tau_k}^{(k,i)},$$

where $X_t^{(k,i)}$ are independent copies of the process X_t. Taking conditional expectation with respect to the values of τ's and taking into account that

$$\mathbb{E}\left\{ \sum_{i=1}^{\xi(k)} X_{t-\tau_k}^{(k,i)} \,\Big|\, \tau_1, \ldots, \tau_{X_s} \right\} \leq m \quad \text{a.s.},$$

by (5.52), we arrive at the following inequalities:

$$\mathbb{E}X_t \leq \mathbb{E} \sum_{k=1}^{X_s} \mathbb{I}\{\tau_k > t\} + \mathbb{E} \sum_{k=1}^{X_s} \mathbb{I}\{\tau_k \leq t\} m \leq \mathbb{E}X_s.$$

Since the function $\mathbb{E}X_t$ is non-increasing, it may be viewed as the tail of an auxiliary probability distribution H on $\mathbb{R}^+$, $\overline{H}(t) := \mathbb{E}X_t$.

We rewrite the equality (5.51) in terms of H:

$$\overline{H}(t) = \overline{G}(t) + m \int_0^t \overline{H}(t-y)G(dy), \tag{5.53}$$

so that $H = (1-m)G + m \cdot H * G$. We iterate this equality to deduce

$$H = (1-m) \sum_{k=0}^{\infty} m^k G^{*(k+1)}. \tag{5.54}$$

Applying the inequality $\overline{H}(t-y) \geq \overline{H}(t)$ to (5.53), we get

$$\overline{H}(t) \geq \overline{G}(t) + m\overline{H}(t)G(t),$$

which implies $\mathbb{E}X_t \geq \overline{G}(t)/(1 - mG(t))$. This shows that, in (5.49), the very left side is not smaller than the very right side.

If G is subexponential, then the representation (5.54) and Theorem 3.37 yields the estimate

$$\mathbb{E}X_t = \overline{H}(t) \sim \frac{\overline{G}(t)}{1-m} \quad \text{as } t \to \infty. \tag{5.55}$$

Now make use of (5.48) in order to obtain an equation for $p_t := \mathbb{P}\{X_t = 0\}$:

$$p_t = \int_0^t G(dy) \sum_{n=0}^{\infty} \mathbb{P}\{\xi^{(1)} = n\} p_{t-y}^n$$

$$= \int_0^t G(dy) \mathbb{E}e^{\xi^{(1)} \log p_{t-y}}. \tag{5.56}$$

Since $\mathbb{E}e^{\xi^{(1)}\log p_{t-y}} \geq 1+\mathbb{E}\xi^{(1)}\log p_{t-y} = 1+m\log p_{t-y}$ and $p_{t-s} \leq p_t$, the equality (5.56) yields

$$p_t \leq G(t)(1+m\log p_t)$$
$$\leq G(t)(1+m(p_t-1)).$$

Therefore,

$$\mathbb{P}\{X_t = 0\} = p_t \leq \frac{G(t)-mG(t)}{1-mG(t)} = 1 - \frac{\overline{G}(t)}{1-mG(t)},$$

which implies the lower bound (5.49).

If G is subexponential, then the decomposition

$$\mathbb{E}X_t = \mathbb{P}\{X_t = 1\} + \mathbb{E}\{X_t; X_t \geq 2\}$$

together with the asymptotics (5.55) and the lower bound (5.49) yield

$$\mathbb{E}\{X_t; X_t \geq 2\} = o(\overline{G}(t)) \quad \text{as } t \to \infty.$$

This observation completes the proof of (5.50). $\square$

5.13 How Do Large Values of M Occur in Standard Cases?

In this section we consider the trajectory of the random walk given that $M > x$, for x large. We assume here that the integrated tail distribution F_I is both subexponential and in the domain of attraction of an extreme-value distribution. We complement Theorem 5.4* from Sect. 5.2 by a limit theorem, as $x \to \infty$, for the distribution of the quadruple that includes the time $\tau(x)$ to exceed level x, position $S_{\tau(x)}$ at this time, position $S_{\tau(x)-1}$ at the prior time, and the trajectory up to it.

Actually, we need the following assumption: there exist a function $e(x) \uparrow \infty$ and a continuous probability distribution F_* on the positive half-line $(0, \infty)$ such that, for any $t > 0$,

$$\frac{\overline{F_I}(x+te(x))}{\overline{F_I}(x)} \to \overline{F}_*(t) \quad \text{as } x \to \infty. \tag{5.57}$$

Let ξ_* be a random variable with distribution F_*. It turns out that (5.57) is satisfied if the integrated tail distribution F_I is in the maximal domain of attraction of an extreme-value distribution. So, in this section, we first formulate and prove a limit theorem (Theorem 5.24) based on the assumption (5.57), and then recall basic concepts and facts from extreme value theory and provide sufficient conditions for Theorem 5.24 to hold.

Introduce stopping times

$$\tau(x) := \min\{n \geq 1 : S_n > x\}$$

(let $\tau(x) = \infty$ if $M \le x$), and

$$\tau_B(x) := \min\{n \ge 1 : B_{n-1}(x) \text{ occures}\},$$

where the events $B_n(x)$ are defined in Theorem 5.4* (let $\tau_B(x) = \infty$ if none of events $B_n(x)$ occurs). Then the following results hold.

Lemma 5.23. *Let* $\mathbb{E}\xi = -a < 0$ *and assume* $F_I \in \mathcal{S}$. *Then*

$$\mathbb{P}\{\tau(x) = \tau_B(x) \mid M > x\} \to 1 \quad as \ x \to \infty.$$

If, in addition, (5.57) holds, then

$$\mathbb{P}\{a\tau(x)/e(x) > t \mid M > x\} \to \overline{F}_*(t) \quad as \ x \to \infty.$$

Proof. Indeed, the first assertion is just rephrasing of the first statement in Theorem 5.4*. Further,

$$\begin{aligned}
\mathbb{P}\{a\tau(x)/e(x) > t \mid M > x\} &\sim \mathbb{P}\{\tau_B(x) > te(x)/a \mid M > x\} \\
&= \frac{\mathbb{P}\{\tau_B(x) > te(x)/a, M > x\}}{\mathbb{P}\{M > x\}} \frac{\mathbb{P}\{\tau_B(x) < \infty, M > x\}}{\mathbb{P}\{\tau_B(x) < \infty, M > x\}} \\
&= \frac{\mathbb{P}\{\tau_B(x) > te(x)/a, M > x\}}{\mathbb{P}\{\tau_B(x) < \infty\}} \mathbb{P}\{\tau_B(x) < \infty \mid M > x\}.
\end{aligned}$$

Therefore, by Theorem 5.4*,

$$\begin{aligned}
\mathbb{P}\{a\tau(x)/e(x) > t \mid M > x\} &\sim \mathbb{P}\{\tau_B(x) > te(x)/a \mid \tau_B(x) < \infty\} \\
&= \frac{\sum_{n > te(x)/a} \mathbb{P}\{B_{n-1}(x)\}}{\sum_{n \ge 1} \mathbb{P}\{B_{n-1}(x)\}},
\end{aligned}$$

since events $B_n(x)$ are disjoint. Hence,

$$\mathbb{P}\{a\tau(x)/e(x) > t \mid M > x\} \sim \frac{\overline{F}_I(x + te(x))}{\overline{F}_I(x)},$$

which converges to $\overline{F}_*(t)$ as $x \to \infty$, by the condition (5.57). $\qquad\square$

From Theorem 5.4* and Lemma 5.23, we deduce the following result.

Theorem 5.24. *Let* $\mathbb{E}\xi = -a < 0$ *and* $F_I \in \mathcal{S}$, *and assume* (5.57) *to hold. Then the distribution of the vector*

$$\left(\frac{a\tau(x)}{e(x)}, \frac{S_{\tau(x)} - x}{e(x)}, \frac{S_{\tau(x)-1}}{e(x)}, \max_{0 \le n \le \tau(x)-1} \frac{|S_n + na|}{\tau(x)} \right), \tag{5.58}$$

conditioned on $\{M > x\}$, weakly converges to the distribution of $(\xi_*, \xi_*', -\xi_*, 0)$ as $x \to \infty$ where both ξ_* and ξ_*' have the same distribution F_* and

$$\mathbb{P}\{\xi_* > s, \xi_*' > t\} = \mathbb{P}\{\xi_* > s + t\} \quad \text{for all } s, t > 0. \tag{5.59}$$

Proof. We have already proved the convergence of the first component in Lemma 5.23. Due to the first assertion of Lemma 5.23, the fourth component conditionally on $\{M > x\}$ converges to 0 if and only if

$$\max_{0 \le n \le \tau_B(x) - 1} \frac{|S_n + na|}{\tau_B(x)}$$

does. We have $|S_n + na| \le n\varepsilon_n + h(x)$ for every $n \le \tau_B(x) - 1$. Then the latter maximum, conditioned on $\{M > x\}$, converges to 0 as $x \to \infty$ if we choose $h(x)$ such that $h(x) = o(e(x))$. In particular, $S_{\tau_B(x)-1}/\tau_B(x)$, conditioned on $\{M > x\}$, converges to $-a$. Applying again the first assertion of Lemma 5.23, we obtain that $S_{\tau(x)-1}/\tau(x)$, conditioned on $\{M > x\}$, converges to $-a$ too. Since

$$\frac{S_{\tau(x)-1}}{e(x)} = \frac{S_{\tau(x)-1}}{\tau(x)} \frac{\tau(x)}{e(x)},$$

the reference to the convergence of the first component completes the proof of the convergence of the third one.

It remains to show the convergence of the second component in (5.58). This follows from upper bound

$$\mathbb{P}\{S_{\tau(x)} > x + te(x) \mid M > x\} \le \mathbb{P}\{M > x + te(x) \mid M > x\} \to \overline{F}_*(t),$$

and from lower bound

$$\mathbb{P}\{S_{\tau(x)} > x + te(x) \mid M > x\} \ge \sum_{n \ge 1} \mathbb{P}\{|S_j + aj| \le j\varepsilon_j + h(x) \text{ for all } j \le n - 1,$$

$$\xi_n > n(a + \varepsilon_n) + h(x) + x + te(x) \mid M > x\}$$

$$\sim (1 + o(1))\mathbb{P}\{M > x + te(x) \mid M > x\}$$

$$\to \overline{F}_*(t) \quad \text{as } x \to \infty.$$

Finally, in order to prove the equality (5.59) write

$$\mathbb{P}\{\tau(x) > se(x)/a, S_{\tau(x)} > x + te(x) \mid M > x\}$$

$$= (1 + o(1))\mathbb{P}\{\tau_B(x) > se(x)/a, S_{\tau_B(x)} > x + te(x) \mid M > x\},$$

and the probability on the right equals to the sum

$$\sum_{n > se(x)/a} \mathbb{P}\{\tau_B(x) = n, S_n > x + te(x) \mid M > x\}$$

$$= \sum_{n > se(x)/a} \mathbb{P}\{B_{n-1}(x), S_n > x + te(x) \mid M > x\}$$

which is not less than

$$\sum_{n>se(x)/a} \mathbb{P}\{|S_j+aj| \le j\varepsilon_j+h(x) \text{ for all } j \le n-1,$$

$$\xi_n > x+te(x)+(n-1)(a+\varepsilon_{n-1})+h(x) \mid M > x\}$$

$$\sim \frac{\sum_{n>se(x)/a} \mathbb{P}\{\xi_n > x+te(x)+(n-1)(a+\varepsilon_{n-1})+h(x)\}}{\mathbb{P}\{M > x\}}$$

$$\sim \frac{\overline{F_I}(x+(s+t)e(x))}{\overline{F_I}(x)} \quad \text{as } x \to \infty,$$

and which is not greater than

$$\sum_{n>se(x)/a} \mathbb{P}\{|S_j+aj| \le j\varepsilon_j+h(x) \text{ for all } j \le n-1,$$

$$\xi_n > x+te(x)+(n-1)(a-\varepsilon_{n-1})-h(x) \mid M > x\}$$

$$\ge \frac{(1+o(1))\sum_{n>se(x)/a} \mathbb{P}\{\xi_n > x+te(x)+(n-1)(a-\varepsilon_{n-1})-h(x)\}}{\mathbb{P}\{M > x\}}$$

$$\sim \frac{\overline{F_I}(x+(s+t)e(x))}{\overline{F_I}(x)} \quad \text{as } x \to \infty.$$

Therefore,

$$\mathbb{P}\{\tau(x) > se(x)/a, S_{\tau(x)} > x+te(x) \mid M > x\} \sim \frac{\overline{F_I}(x+(s+t)e(x))}{\overline{F_I}(x)}$$

$$\to \overline{F}_*(s+t) \quad \text{as } x \to \infty,$$

which is equivalent to (5.59). □

Now we recall basic concepts and facts from extreme value theory which provide sufficient conditions for (5.57) and Theorem 5.24 to hold.

For independent identically distributed random variables $\eta_1, \eta_2, \ldots$ with distribution G and for non-degenerate distribution H, we say G *belongs to the maximum domain of attraction of* H, if there exist a positive sequence c_n and a real-valued sequence d_n such that the distribution of $c_n(\max(\eta_1,\ldots,\eta_n)-d_n)$ weakly converges to H as $x \to \infty$. Then H is called *an extreme value distribution*.

In the case where G has right-unbounded support, $G(x) < 1$ for all x, there are only two classes of the extreme value distributions, Frechet distribution Φ_α, $\alpha \in (0,\infty)$,

$$H(x) = \Phi_\alpha(x) = e^{-x^{-\alpha}}, \quad x \ge 0,$$

and Gumbel distribution Λ,

$$H(x) = \Lambda(x) = e^{-e^{-x}}.$$

The following two results may be found in Sect. 8.13.2 of [9].

Theorem 5.25. *Assume that $G(x) < 1$ for all x and $\alpha > 0$. Then the following assertions are equivalent:*

(i) *G belongs to the maximum domain of attraction of Φ_α.*
(ii) *G is regularly varying distribution with index $-\alpha$.*
(iii) *There exists a positive measurable function $e(x)$ such that, for any $t > 0$,*

$$\frac{\overline{G}(x+te(x))}{\overline{G}(x)} \to \frac{1}{(1+t/\alpha)^\alpha} \quad as\ x \to \infty.$$

Theorem 5.26. *Assume that $G(x) < 1$ for all x. Then the following assertions are equivalent:*

(i) *G belongs to the maximum domain of attraction of Λ.*
(ii) *The inverse function $R^{(-1)}(x)$ of the hazard function $R(x) = -\log\overline{G}(x)$ is such that, for every fixed $y > 0$, $R^{(-1)}(x+y) - R^{(-1)}(x) \sim u\ell(e^x)$ as $x \to \infty$, for some slowly varying function ℓ.*
(iii) *There exists a positive measurable function $e(x)$ such that, for any $t > 0$,*

$$\frac{\overline{G}(x+te(x))}{\overline{G}(x)} \to e^{-t} \quad as\ x \to \infty.$$

In both cases, the function $e(x)$ can be chosen as $e(x) \sim \overline{G}_I(x)/\overline{G}(x)$.

For example, if the tail of G is equivalent to that of Weibull distribution, $\overline{G}(x) \sim e^{-x^\beta}$ as $x \to \infty$, $\beta \in (0,1)$, then G belongs to the maximum domain of attraction of Λ and $e(x) \sim x^{1-\alpha}/\alpha$.

If instead the tail of G is equivalent to that of the lognormal distribution with parameters μ and σ^2, then G belongs to the maximum domain of attraction of Λ and $e(x) \sim \sigma^2 x/(\log x - \mu)$.

5.14 Comments

Theorem 5.2 was proved for regularly varying distributions by Callaert and Cohen in [12] and by Cohen in [17]. For dominated-varying distributions, it was proved by A. Borovkov in [10, Sect. 22]. In its present form, it was proved by Veraverbeke in [53] and by Embrechts, Goldie and Veraverbeke in [23]. The proof given here follows an idea of Zachary [54].

Theorem 5.3 was proved by Korshunov [36] under slightly different condition, see also Denisov, Foss and Korshunov [18, Corollary 4].

Lemma 5.9 is due to A. Borovkov [10, Sect. 22, Theorem 10].

A proof of the converse part of Theorem 5.12 is given by Korshunov [35]. It was proved by Embrechts and Veraverbeke in [25, Corollary 6.1] and by Pakes in [44, Theorem 1] for the case where ξ_1 is a difference of two independent random variables $\xi_1 = \eta - \zeta$, where ζ has an exponential distribution and $\eta \geq 0$.

Asmussen, Kalashnikov, Konstantinides, Klüppelberg and Tsitsiashvili [5] proved that if $F \in \mathcal{S}^*$ then, for any fixed $T \in (0, \infty)$, $\mathbb{P}\{M \in x + \Delta\} \sim T\overline{F}(x)/a$ (if the distribution F is lattice, then x and T should be restricted to values of the lattice span). In the lattice case, it was proved earlier by Bertoin and Doney [8]. They also sketched a proof for non-lattice distributions. The current version is due to Asmussen, Foss and Korshunov [4]. Density of M was studied in Asmussen, Kalashnikov, Konstantinides, Klüppelberg and Tsitsiashvili [5], in Asmussen, Foss and Korshunov [4], and in Korshunov [37].

The idea of using the inequality $B_r^{*n}(x) \leq B_r^n(x)$ for proving the lower bound given in Sect. 5.9 goes back to Kalashnikov and Tsitsiashvili [31, Theorem 7]. A different proof for that may be found in Korolev, Bening and Shorgin [34], see Theorem 8.7.2 there.

The results for subcritical branching processes under subexponential assumptions was first proved by Chistyakov [13] and further by Chover, Ney and Wainger [15]; see also Athreya and Ney [7].

The limit theorem from Sect. 5.13 was obtained by Asmussen and Klüppelberg in [6] by a different approach, see also Asmussen and Foss [3].

5.15 Problems

5.1. Let ξ_1 have negative mean and heavy-tailed distribution. Prove the maximum M has a heavy-tailed distribution too.

5.2. Let ξ_1 have negative mean and a light-tailed distribution. Prove the maximum M has a light-tailed distribution too.

Hint: Estimate from above the exponential moments of M via those of S_n.

5.3. Suppose that $\xi_1, \ldots, \xi_n$ are independent random variables with common exponential distribution. Find the asymptotic behaviour of the tail probability of the maximum of sums

$$\max_{k \leq n} (\xi_1^\alpha + \ldots + \xi_k^\alpha)$$

for (i) $\alpha > 1$; (ii) $\alpha < 0$.

5.4. Suppose that $\xi_1, \xi_2, \ldots,$ are independent random variables which are uniformly distributed in the interval $[0, 1]$. For $0 < \alpha < 1$ and $\varepsilon > 0$, find the asymptotic behaviour of the distribution density of the maximum of the sums

$$\max_{n \geq 1} \left(\frac{1}{\xi_1^\alpha} + \ldots + \frac{1}{\xi_n^\alpha} - n \left(\frac{1}{1 - \alpha} + \varepsilon \right) \right).$$

5.5. Let $X(t)$ be a compound Poisson process with a subexponential distribution for a typical jump. For every T, find the tail asymptotics for the distribution of the supremum

$$\sup_{0 \leq t \leq T} X(t)$$

in terms of the jump distribution.

5.6. Suppose $X(t)$ is a compound Poisson process with a negative drift $-a$ and with a jump distribution F. Find the tail asymptotics for the distribution of the supremum

$$\sup_{t \geq 0} X(t)$$

provided F is regularly varying at infinity with index $\alpha > 0$. Find what power moments of this supremum are finite and which are infinite.

5.7. *Moving overages.* Suppose $\eta_0, \eta_1, \eta_2, \ldots$ are independent identically distributed random variables with common distribution F and zero mean. Let $\xi_n := \eta_n + \eta_{n-1} - a$ where $a > 0$. Define $S_0 := 0$, $S_n := \xi_1 + \ldots + \xi_n$ and $M := \max\{S_n, n \geq 0\}$. Prove that M is finite with probability 1 and that:

(i) If the integrated tail distribution F_I is long-tailed, then

$$\liminf_{x \to \infty} \frac{\mathbb{P}\{M > x\}}{2\overline{F}_I(x/2)} \geq \frac{1}{a}.$$

(ii) If the integrated tail distribution F_I is subexponential, then

$$\mathbb{P}\{M > x\} \sim 2\overline{F}_I(x/2)/a \quad \text{as } x \to \infty.$$

5.8. In the conditions of the previous problem, let $\xi_n := \eta_n - \eta_{n-1} - a$ where $a > 0$, and the η's be bounded from below. Then what is the tail asymptotics for the distribution of M? How essential is the condition that the η's are bounded from below?

5.9. Suppose $k \geq 1$ and $c_i \geq 0$ for all $i = 0, \ldots, k$. Generalise the result of Problem 7 onto the moving overage

$$\xi_n := c_0 \eta_n + c_1 \eta_{n-1} + \ldots + c_k \eta_{n-k} - a.$$

5.10. *Maximum of a skeleton.* Suppose $\xi_1, \xi_2, \ldots$ are independent identically distributed random variables with common distribution F and mean $-a < 0$. Fix $k \geq 2$. Define $S_0 := 0$, $S_n := \xi_1 + \ldots + \xi_n$ and

$$M(k) := \max\{S_{kn}, n \geq 0\}.$$

Prove that if the integrated tail distribution F_I is subexponential, then

$$\mathbb{P}\{M(k) > x\} \sim \frac{\overline{F}_I(x)}{a} \quad \text{as } x \to \infty.$$

5.11. Suppose $\xi_1, \xi_2, \ldots$ are independent random variables with common subexponential distribution F. Let τ be a counting random variable which has a light-tailed distribution and doesn't depend on the ξ's. Prove

$$\mathbb{P}\{\max_{n \leq \tau} S_n > x\} \sim \mathbb{E}\tau \overline{F}(x) \quad \text{as } x \to \infty.$$

5.12. *Excess process.* In the conditions of Problem 2.26, provided the distribution F of the jump at state 1 is long-tailed, prove

$$\mathbb{P}\{X_n > x \mid X_0 = 1\} \sim \frac{1}{\mathbb{E}\{X_1 \mid X_0 = 1\}} \int_x^{x+n} \overline{F}(y)\,dy$$

as $n, x \to \infty$.

5.13. *Periodicity in time.* Let F be a distribution on $\mathbb{R}$. Suppose $\{\xi_n^{(1)}\}$ and $\{\xi_n^{(2)}\}$ are two independent sequences of independent random variables such that the $\xi^{(1)}$'s have common distribution $F^{(1)}$ with mean $a^{(1)}$ while the $\xi^{(2)}$'s have common distribution $F^{(2)}$ with mean $a^{(2)}$. Suppose that $\overline{F^{(1)}}(x) \sim c_1\overline{F}(x)$ and $\overline{F^{(2)}}(x) \sim c_2\overline{F}(x)$ as $x \to \infty$. Suppose also that $a^{(1)} + a^{(2)} < 0$ and $c^{(1)} + c^{(2)} > 0$. Put $\xi_{2n-1} := \xi_n^{(1)}$, $\xi_{2n} := \xi_n^{(2)}$ for $n \geq 1$ and $S_0 := 0$, $S_n := \xi_1 + \ldots + \xi_n$. Given that the integrated tail distribution F_I is subexponential, prove that

$$\mathbb{P}\left\{\max_{n \geq 0} S_n > x\right\} \sim \frac{(c^{(1)} + c^{(2)})\overline{F_I}(x)}{|a^{(1)} + a^{(2)}|} \quad \text{as } x \to \infty.$$

Also find the finite time horizon asymptotics for $\mathbb{P}\{\max_{0 \leq i \leq n} S_i > x\}$ as $x \to \infty$.

5.14. *Perturbation in space.* Let F be a distribution on $\{-1,0,1,2,3,\ldots\}$ with negative mean $-a < 0$. Suppose X_n, $n \geq 0$, is a time-homogeneous Markov chain on $\mathbb{Z}^+$ such that, for every $i \geq 1$, the distribution of the jump $X_1 - X_0$, conditioned to $X_0 = i$, is F. Assume that the jump size from the state 0 has an arbitrary distribution which is bounded from above. Assume that the integrated tail distribution F_I is subexponential. Prove the tail of the invariant measure π is equivalent to

$$\sum_{i=x}^{\infty} \pi_i \sim \frac{c}{a}\overline{F_I}(x) \quad \text{as } x \to \infty,$$

where the constant c is equal to

$$c = 1 - \pi_0 = \frac{\mathbb{E}\{X_1 \mid X_0 = 0\}}{a + \mathbb{E}\{X_1 \mid X_0 = 0\}}.$$

Also find the asymptotics for $\mathbb{P}\{X_n > x\}$ as $n, x \to \infty$.

5.15. *Perturbation in space–continuation.* Let F be a distribution on $\{-1,0,1,2,3,\ldots\}$ with negative mean $-a < 0$. Let i_0 be a positive integer. Suppose X_n, $n \geq 0$, is a time-homogeneous Markov chain on $\mathbb{Z}^+$ such that, for every $i \geq i_0$, the distribution of the jump $X_1 - X_0$ conditioned to $X_0 = i$ is F. Assume further that all distributions of the jumps from the states $0, \ldots, i_0 - 1$ are bounded from above. Given the integrated tail distribution F_I is subexponential, prove the tail of the invariant measure π is equivalent to

$$\sum_{i=x}^{\infty} \pi_i \sim \frac{c}{a}\overline{F_I}(x) \quad \text{as } x \to \infty,$$

where the constant c is equal to

$$c = 1 - \sum_{j=i_0}^{i_0-1} \pi_j.$$

Answers to Problems

Chapter 2

2.9. Yes, it can. Consider, for example, probability space $\Omega = [0,1]$ with Borel sigma-algebra and Lebesgue measure, and let $\xi(\omega) = \frac{1}{\omega}\mathbb{I}\{\omega < 1/2\}$ and $\eta(\omega) = \xi(1-\omega)$. Then $\min(\xi,\eta) = 0$.

2.14. Exponential distribution, for example.

2.18. (i) $\alpha > 2$.

2.19. $n \geq 3$. Yes, the product is long-tailed.

2.20. $\beta < n$.

2.21. (ii) The same result holds for any distribution with negative mean.

2.26. (i) *Solution.* Denote by τ the first return time to the state 1, that is, $\tau = \min(n \geq 1 : X_n = 1)$ given $X_0 = 1$. Then $\tau = n$ if and only if $X_1 = n$, so that $\mathbb{P}\{\tau = n\} = F\{n-1\}$ and $\mathbb{E}\tau = \sum_{n \geq 1} nF\{n-1\}$. Therefore, the Markov chain is positive recurrent if and only if F has finite mean.

Let $\{\pi_i\}_{i \geq 1}$ be the stationary distribution. For any $i \geq 1$,

$$\pi_i = \pi_{i+1} + \pi_1 F\{i-1\}$$
$$= \pi_{i+2} + \pi_1 F\{i\} + \pi_1 F\{i-1\}$$
$$= \dots$$
$$= \pi_1 F[i-1, \infty).$$

Then

$$\sum_{i \geq 1} \pi_i = \pi_1 \sum_{i \geq 1} \sum_{j \geq i-1} F\{j\}$$
$$= \pi_1 \sum_{j \geq 0} (j+1)F\{j\},$$

which implies that the invariant distribution is given by the residual distribution F_r.

S. Foss et al., *An Introduction to Heavy-Tailed and Subexponential Distributions*, Springer Series in Operations Research and Financial Engineering, DOI 10.1007/978-1-4614-7101-1, © Springer Science+Business Media New York 2013

Chapter 3

3.8. Let the ξ's have exponential distribution with parameter λ.

(i) $n_1 e^{-\lambda x^{1/\beta_1}}$ where $\beta_1 := \max \alpha_i$ and n_1 is the number of α_i that are equal to β_1;

(ii) $n_2 \lambda x^{1/\beta_2}$ where $\beta_2 := \min \alpha_i$ and n_2 is the number of α_i that are equal to β_2;

(iii) $n_1 e^{-\lambda x^{1/\beta_1}}$ if $\beta_2 > 1$ and $n_2 \lambda x^{1/\beta_2}$ if $\beta_2 < 0$.

3.9. $\alpha < 1 - \beta$.

3.10. For all positive values.

3.12. $\mathbb{P}\{S_n > x\} \le n\mathbb{P}\{\xi_1 > x - 1\}$.

3.13. $\lambda t \overline{F}(x)$ where λ is the intensity and F is the jump distribution.

3.15. Proportional to $\overline{F}(x)$ with the following coefficients:

(i) $\mathbb{P}\{X_0 = 1\}(2p_{11} + p_{12}(1+c)) + \mathbb{P}\{X_0 = 2\}(p_{22}2c + p_{21}(1+c))$;

(ii) $\mathbb{P}\{X_0 = 1\}\left[1 + p_{11} + \sum_{j=0}^{\infty} p_{12}p_{21}p_{22}^j((j+1)c+1)\right] + \mathbb{P}\{X_0 = 2\}\left[c + p_{22}c + \sum_{j=0}^{\infty} p_{21}p_{11}^j p_{12}(j+1+c)\right]$;

(iii) $\mathbb{P}\{X_0 = 1\}\left[1 + kp_{11} + k\sum_{j=0}^{\infty} p_{12}p_{22}^j p_{21}((j+1)c+1)\right] + \mathbb{P}\{X_0 = 2\}\left[c + kp_{22}c + k\sum_{j=0}^{\infty} p_{21}p_{11}^j p_{12}(j+1+c)\right]$.

3.16. (i) $H(x) = x$; (ii) $H(x) = x/\log x$; (iii) $H(x) = x^{1-\beta}$; (iv) $H(x) = \log x$.

3.19. (i) $\frac{1}{x\log x}$ and $\frac{2}{x\log x}$; (ii) $\frac{e^{-\sqrt{x}}}{\sqrt{x}\log x}$ and $\frac{2e^{-\sqrt{x}}}{\sqrt{x}\log x}$; (iii) $\frac{c_1}{\log x}$ and $\frac{c_2}{\log x}$ where $c_2 < 2c_1$.

Solution. The tail distribution function of ξ_i is equal to

$$\overline{F}(x) = \int_0^1 e^{-x^y} dy = \frac{1}{\log x}\int_1^x \frac{e^{-u}}{u} du \sim \frac{c_1}{\log x} \qquad \text{as } x \to \infty,$$

where

$$c_1 := \int_1^\infty \frac{e^{-u}}{u} du.$$

Further,

$$\mathbb{P}\{\xi_1 + \xi_2 > x\} = \mathbb{P}\{\xi_1 > x\} + \mathbb{P}\{\xi_2 > x\} - \mathbb{P}\{\xi_1 > x, \xi_2 > x\}$$
$$+ \mathbb{P}\{\xi_1 \le x, \xi_2 \le x, \xi_1 + \xi_2 > x\}.$$

By the conditional independence,

$$\mathbb{P}\{\xi_1 > x, \xi_2 > x\} = \mathbb{E}\mathbb{P}\{\xi_1 > x, \xi_2 > x \mid \eta\}$$
$$= \int_0^1 e^{-2x^t} dt \sim \frac{1}{\log x}\int_2^\infty \frac{e^{-u}}{u} du \qquad \text{as } x \to \infty.$$

It is left to prove that

$$\mathbb{P}\{\xi_1 \le x, \xi_2 \le x, \xi_1 + \xi_2 > x\} = o(\overline{F}(x)) \qquad \text{as } x \to \infty. \tag{5.60}$$

Indeed, this probability is equal to

$$\int_0^1 dt \int_0^x tu^{t-1}e^{-u^t}du \int_{x-u}^x tv^{t-1}e^{-v^t}dv$$
$$= \int_0^1 x^{2t}dt \int_0^1 \int_{1-z}^1 t^2 y^{t-1}z^{t-1}e^{-x^t(y^t+z^t)}dydz.$$

Since $y^t + z^t \geq 1$, the latter term is not greater than

$$\int_0^1 x^{2t}e^{-x^t}dt \int_0^1 \int_{1-z}^1 t^2 y^{t-1}z^{t-1}dydz = \int_0^1 x^{2t}e^{-x^t}J(t)dt,$$

where $J(t) = \mathbb{P}\{Y_1^{1/t} + Y_2^{1/t} > 1\}$ and Y_1, Y_2 are independent random variables both uniformly distributed in the interval $[0,1]$. As $t \downarrow 0$, $J(t) \downarrow 0$. Hence, (5.60) follows from the estimates, for any $\varepsilon > 0$,

$$\int_0^1 x^{2t}e^{-x^t}J(t)dt = \left(\int_0^\varepsilon + \int_\varepsilon^1\right)x^{2t}e^{-x^t}J(t)dt$$
$$\leq J(\varepsilon)\int_0^\varepsilon x^{2t}e^{-x^t}dt + \int_\varepsilon^1 x^{2t}e^{-x^t}dt$$
$$\leq \frac{J(\varepsilon)}{\log x}\int_1^\infty te^{-t}dt + x^2 e^{-x^\varepsilon}.$$

Thus, the answer is:

$$\mathbb{P}\{\xi_1 + \xi_2 > x\} \sim \frac{c_2}{\log x} \quad \text{as } x \to \infty,$$

where

$$c_2 := \left(2\int_1^\infty - \int_2^\infty\right)\frac{e^{-u}}{u}du.$$

3.20. Proportional to $\overline{F}(x)$ with the following coefficient:

(i) $2\mathbb{E}e^{-\eta_1}$;   (ii) $(1+\mathbb{E}e^{-\eta_1})\mathbb{E}e^{-\eta_2}$.

3.22. All the integrated tail distributions $F_{i,I}$ are tail-proportional to a reference subexponential distribution.

Chapter 4

4.6. Yes, it is.

    4.11. Let the ξ's have exponential distribution with parameter λ.

(i) $n_1\lambda x^{1/\beta_1-1}e^{-\lambda x^{1/\beta_1}}$ where $\beta_1 := \max \alpha_i$ and n_1 is the number of α_i that are equal to β_1;

(ii) $n_2\lambda x^{1/\beta_2}/|\beta_2|$ where $\beta_2 := \min \alpha_i$ and n_2 is the number of α_i that are equal to β_2;

(iii) $n_1\lambda x^{1/\beta_1-1}e^{-\lambda x^{1/\beta_1}}$ if $\beta_2 > 1$ and $n_2\lambda x^{1/\beta_2}/|\beta_2|$ if $\beta_2 < 0$.

4.12. $\sum_n \frac{n\mathbb{P}\{\tau=n\}}{\pi(n+x^2)}$.

4.14. $\lambda t f(x)$.

4.19.

(i) $p_{i,i+1} = p - c/i + o(1/i)$ and $p_{i,i-1} = p + c/i + o(1/i)$ as $i \to \infty$ where $p < 1/2$ and $c > p/2$;

(ii) $p_{i,i+1} = p - c/i^\beta + o(1/i^\beta)$ and $p_{i,i-1} = p + c/i^\beta + o(1/i^\beta)$ as $i \to \infty$ where $p < 1/2, c > 0$ and $0 < \beta < 1$.

4.20. *Solution.* It follows from the solution to Problem 2.26 that the invariant distribution for the Markov chain coincides with the residual distribution F_r. Then the result follows from Theorem 4.32 with $T = 1$.

4.21. All the distributions F_i are tail equivalent to some distribution from $\mathcal{S}^*$.

4.22. $\frac{(1-p)^2}{(2\pi)^{d/2}\sqrt{\det B}}$.

Chapter 5

5.3. Let the ξ's have exponential distribution with parameter λ.

(i) $ne^{-\lambda x^{1/\alpha}}$;   (ii) $n\lambda x^{1/\alpha}$.

5.4. $\frac{\alpha}{\varepsilon(1-\alpha)}x^{1-1/\alpha}$.

5.5. It is proportional to the tail of the jump distribution with coefficient λT where λ is the intensity of the jumps.

5.6. It is proportional to the tail $\overline{F}_I$ with coefficient λ/a where λ is the intensity of the jumps.•

5.8. $\overline{F}_I(x)/a$. If the left tail of the distribution of η is much heavier than the right tail, then the asymptotic tail behaviour of M will be determined by the left tail of η.

5.9. $\mathbb{P}\{M > x\} \sim c\overline{F}_I(x/c)/a$ where $c := c_0 + \ldots + c_k$.

5.12. *Solution.* We observe that

$$\mathbb{P}\{X_n > x \mid X_0 = 1\} = \sum_{k=1}^{n} \mathbb{P}\{X_{n-k} = 1, X_{n-k+1} \geq 2, \ldots, X_{n-1} \geq 2, X_n > x \mid X_0 = 1\}.$$

Then, by the Markov property,

$$\mathbb{P}\{X_n > x \mid X_0 = 1\} = \sum_{k=1}^{n} \mathbb{P}\{X_{n-k} = 1 \mid X_0 = 1\}$$

$$\times \mathbb{P}\{X_k > x, X_{k-1} \geq 2, \ldots, X_1 \geq 2 \mid X_0 = 1\}$$

$$= \sum_{k=1}^{n} \mathbb{P}\{X_{n-k} = 1 \mid X_0 = 1\}F[x+k, \infty).$$

By the ergodic theorem for Markov chains,

$$\mathbb{P}\{X_n = 1 \mid X_0 = 1\} \to 1/\mathbb{E}\{\tau \mid X_0 = 1\} \quad \text{as } n \to \infty,$$

where $\tau = \min\{n \geq 1 : X_n = 1\}$. Taking also into account that the distribution F is long-tailed, we deduce that, as $n, x \to \infty$,

$$\mathbb{P}\{X_n > x \mid X_0 = 1\} \sim \frac{1}{\mathbb{E}\{\tau \mid X_0 = 1\}} \sum_{k=1}^{n} F[x+k, \infty)$$

$$\sim \frac{1}{\mathbb{E}\{\tau \mid X_0 = 1\}} \int_{x}^{x+n} \overline{F}(y) dy.$$

Since $\mathbb{E}\{\tau \mid X_0 = 1\} = \mathbb{E}\{X_1 \mid X_0 = 1\}$, the result follows.

5.13. The answers differ for even and odd valued of n. If n is even number, $n = 2k$, then

$$\mathbb{P}\{\max_{0 \leq i \leq n} S_i > x\} \sim \frac{(c^{(1)} + c^{(2)})}{|a^{(1)} + a^{(2)}|} \int_{0}^{k|a^{(1)} + a^{(2)}|} \overline{F}(x+y) dy$$

as $x \to \infty$ uniformly in $k \geq 1$. If n is odd, $n = 2k + 1$, then

$$\mathbb{P}\{\max_{0 \leq i \leq n} S_i > x\} \sim \frac{(c^{(1)} + c^{(2)})}{|a^{(1)} + a^{(2)}|} \int_{0}^{k|a^{(1)} + a^{(2)}|} \overline{F}(x+y) dy + c^{(2)} \overline{F}(x).$$

5.14. $\frac{c}{a} \int_{0}^{n|a|} \overline{F}(x+y) dy$ as $n, x \to \infty$.

References

[1] Asmussen, S.: Applied Probability and Queues, 2nd edn. Springer, New York (2003)

[2] Asmussen, S.: Ruin Probabilities. World Scientific, Singapore (2000)

[3] Asmussen, S., Foss, S.: On exceedance times for some processes with dependent increments. J. Appl. Probab. **51** (2014)

[4] Asmussen, S., Foss, S., Korshunov, D.: Asymptotics for sums of random variables with local subexponential behaviour. J. Theor. Probab. **16**, 489–518 (2003)

[5] Asmussen, S., Kalashnikov, V., Konstantinides, D., Klüppelberg, C., Tsitsiashvili, G.: A local limit theorem for random walk maxima with heavy tails. Stat. Probab. Lett. **56**, 399–404 (2002)

[6] Asmussen, S., Klüppelberg, C.: Large deviations results for subexponential tails, with applications to insurance risk. Stoch. Proc. Appl. **64**, 103–125 (1996)

[7] Athreya, K., Ney, P.: Branching Processes. Springer, Berlin (1972)

[8] Bertoin, J., Doney, R.A.: On the local behaviour of ladder height distributions. J. Appl. Probab. **31**, 816–821 (1994)

[9] Bingham, N.H., Goldie, C.M., Teugels, J.L.: Regular Variation. Cambridge University Press, Cambridge (1987)

[10] Borovkov, A.A.: Stochastic Processes in Queueing Theory. Springer, Berlin (1976)

[11] Borovkov, A.A., Borovkov, K.A.: Asymptotic Analysis of Random Walks. Heavy-Tailed Distributions. Cambridge University Press, Cambridge (2008)

[12] Callaert, H., Cohen, J.W.: A lemma on regular variation of a transient renewal function. Z. Wahrscheinlichkeitstheorie verw. Gebiete **24**, 275–278 (1972)

[13] Chistyakov, V.P.: A theorem on sums of independent positive random variables and its application to branching random processes. Theor. Probab. Appl. **9**, 640–648 (1964)

[14] Chover, J., Ney, P., Wainger, S.: Functions of probability measures. J. d'Analyse Mathématique **26**, 255–302 (1973)

[15] Chover, J., Ney, P., Wainger, S.: Degeneracy properties of subcritical branching processes. Ann. Probab. **1**, 663–673 (1973)

[16] Cline, D.: Convolutions of distributions with exponential and subexponential tails. J. Aust. Math. Soc. **43**, 347–365 (1987)

[17] Cohen, J.W.: Some results on regular variation for distributions in queueing and fluctuation theory. J. Appl. Probab. **10**, 343–353 (1973)

[18] Denisov, D., Foss, S., Korshunov, D.: Tail asymptotics for the supremum of a random walk when the mean is not finite. Queueing Syst. **46**, 15–33 (2004)

[19] Denisov, D., Foss, S., Korshunov, D.: On lower limits and equivalences for distribution tails of randomly stopped sums. Bernoulli **14**, 391–404 (2008)

[20] Denisov, D., Foss, S., Korshunov, D.: Lower limits for distribution tails of randomly stopped sums. Theor. Probab. Appl. **52**, 690–699 (2008)

[21] Embrechts, P., Goldie, C.M.: On closure and factorization theorems for subexponential and related distributions. J. Aust. Math. Soc. Ser. A **29**, 243–256 (1980)

[22] Embrechts, P., Goldie, C.M.: On convolution tails. Stoch. Proc. Appl. **13**, 263–278 (1982)

[23] Embrechts, P., Goldie, C.M., Veraverbeke, N.: Subexponentiality and infinite divisibility. Z. Wahrscheinlichkeitstheorie verw. Gebiete **49**, 335–347 (1979)

[24] Embrechts, P., Klüppelberg, C., Mikosch, T.: Modelling Extremal Events for Insurance and Finance. Springer, Berlin (1997)

[25] Embrechts, P., Veraverbeke, N.: Estimates for the probability of ruin with special emphasis on the possibility of large claims. Insur. Math. Econ. **1**, 55–72 (1982)

[26] Feller, W.: An Introduction to Probability Theory and Its Applications, vol. 2. Wiley, New York (1971)

[27] Foss, S., Korshunov, D.: Lower limits and equivalences for convolution tails. Ann. Probab. **35**, 366–383 (2007)

[28] Frisch, U., Sornette, D.: Extreme deviations and applications. J. Phys. I France **7**, 1155–1171 (1997)

[29] Hallinan, A.J.: A review of the Weibull distribution. J. Qual. Tech. **25**, 85–93 (1993)

[30] Kalashnikov, V.: Geometric Sums: Bounds for Rare Events with Applications. Kluwer Academic Publishers, Dordrecht (1997)

[31] Kalashnikov, V., Tsitsiashvili, G.: Tails of waiting times and their bounds. Queueing Syst. **32**, 257–283 (1999)

[32] Klüppelberg, C.: Subexponential distributions and integrated tails. J. Appl. Probab. **25**, 132–141 (1988)

[33] Klüppelberg, C.: Subexponential distributions and characterization of related classes. Probab. Theor. Rel. Fields **82**, 259–269 (1989)

[34] Korolev, V.Yu., Bening, V.E., Shorgin, S.Ya.: Mathematical foundations of risk theory. Fizmatlit, Moscow (in Russian) (2007)

[35] Korshunov, D.: On distribution tail of the maximum of a random walk. Stoch. Proc. Appl. **72**, 97–103 (1997)

[36] Korshunov, D.: Large-deviation probabilities for maxima of sums of independent random variables with negative mean and subexponential distribution. Theor. Probab. Appl. **46**, 355–366 (2002)

[37] Korshunov, D.: On the distribution density of the supremum of a random walk in the subexponential case. Siberian Math. J. **47**, 1060–1065 (2006)

[38] Korshunov, D.: How to measure the accuracy of the subexponential approximation for the stationary single server queue. Queueing Syst. **68**, 261–266 (2011)

[39] Laherrère, J., Sornette, D.: Streched exponential distributions in nature and economy: "fat tails" with characteristic scales. Eur. Phys. J. B **2**, 525–539 (1998)

[40] Leslie, J.R.: On the non-closure under convolution of the subexponential family. J. Appl. Probab. **26**, 58–66 (1989)

[41] Lindley, D. V.: The theory of queues with a single server. Proc. Cambridge Philos. Soc. **8**, 277–289 (1952)

[42] Malevergne, Y., Sornette, D.: Extreme Financial Risks: From Dependence to Risk Management. Springer, Heidelberg (2006)

[43] Metzler, R., Klafter, J.: The random walk's guide to anomalous diffusion: a fractional dynamics approach. Phys. Rep. **339**, 1–77 (2000)

[44] Pakes, A.G.: On the tails of waiting-time distributions. J. Appl. Probab. **12**, 555–564 (1975)

[45] Pitman, E.J.G.: Subexponential distribution functions. J. Aust. Math. Soc. Ser. A **29**, 337–347 (1980)

[46] Rogozin, B.A.: On the constant in the definition of subexponential distributions. Theor. Probab. Appl. **44**, 409–412 (2000)

[47] Rolski, T., Schmidli, H., Schmidt, V., Teugels, J.: Stochastic Processes for Insurance and Finance. Wiley, Chichester (1998)

[48] Rudin, W.: Limits of ratios of tails of measures. Ann. Probab. **1**, 982–994 (1973)

[49] Sgibnev, M.S.: Banach algebras of functions that have identical asymptotic behaviour at infinity. Siberian Math. J. **22**, 179–187 (1981)

[50] Seneta, E.: Regularly Varying Functions, Springer, Berlin (1976)

[51] Sornette, D.: Critical Phenomena in Natural Sciences, 2nd edn. Springer, Berlin (2004)

[52] Teugels, J.L.: The class of subexponential distributions. Ann. Probab. **3**, 1000–1011 (1975)

[53] Veraverbeke, N.: Asymptotic behaviour of Wiener-Hopf factors of a random walk. Stoch. Proc. Appl. **5**, 27–37 (1977)

[54] Zachary, S.: A note on Veraverbeke's theorem. Queueing Syst. **46**, 9–14 (2004)

Index

S. Foss et al., *An Introduction to Heavy-Tailed and Subexponential Distributions*, 155
Springer Series in Operations Research and Financial Engineering,
DOI 10.1007/978-1-4614-7101-1, © Springer Science+Business Media New York 2013

Printed in the United States
By Bookmasters